2.95

The Random House Dictionary of New Information Technology

The Random House Dictionary of New Information Technology

A.J. Meadows
M. Gordon
A. Singleton

Vintage Books
A Division of Random House
New York

First Vintage Books Edition, January 1983
Copyright © 1982 by A.J. Meadows and
Kogan Page Ltd.

Library of Congress Cataloging in Publication Data
Dictionary of new information technology.
The Random House dictionary of new information technology.
Originally published: Dictionary of new information technology.
London : Kogan Page ; New York : Nichols Pub. Co., 1982.
1. Electronic data processing—Dictionaries.
2. Telecommunication—Dictionaries.
I. Meadows, A. J. (Arthur Jack)
II. Gordon, M.
III. Singleton, A.
IV. Title.
QA76.15.D527   1983   001.5′03′21   82-40026
ISBN 0-394-71202-1

Introduction

Information technology, as a general term, was introduced only recently, but forms of information technology – for example, the telephone – have been familiar for years. What distinguishes new information technology from the older types is the way it combines a variety of communication channels with the information-handling capabilities of computers. New (mainly electronic) methods for dealing with the generation, transmission and reception of information are proliferating rapidly. At the same time, methods of communication which have traditionally progressed separately (for example, telephone and television) are now being drawn together. The rate of development is such that even active participants in the communication system find it hard to keep up with progress outside their own particular sphere. Members of the general public are often, quite naturally, totally confused by what is happening. By way of introduction to this dictionary, it is therefore worthwhile describing briefly some of the techniques that are involved in new information technology. This should also serve to give some indication of the dictionary's intended scope. (Italicized words indicate that the term is discussed, usually at some length, in the body of the dictionary.)

We can start with *computers*, since these represent one of the basic elements of the new technology. Reasonably powerful computers have been decreasing rapidly both in size and cost over the past decade. With the advent of the *microprocessor*, computers have been developed which can sit on a desk top, where they can be used for generating and transmitting information. A typical example of this use is in handling the vast quantities of text generated in the modern world. Specialized computers (called *word processors*) that can automatically produce a range of letters and documents are beginning to revolutionize office work. They are spreading into every field – for example, publishing – that involves the generation and handling of text. The *output* from such computers can be produced as traditional print-on-paper; but, since the computer holds the material in electronic form, it can be transmitted with equal ease through other communication channels, eg a telephone *network*, to a distant receiver. A letter can thus by-pass the traditional mail system, to be delivered by *electronic mail*.

This is just one example of information handling with the new technology. Methods both for the *input* of information to, and the output of information from a system are diversifying at a bewildering speed. Most information is still converted to electronic form via a *keyboard*, as with a typewriter. But it is already possible to talk to a computer, and for some speech to be accepted directly (*voice input*). It is confidently expected that current limitations, eg in the computer's vocabulary, will at least partly disappear during the 1980s.

At the output stage, computerized systems can deal with so much information that new methods must be found for storing it all. The development of the *videodisc* may make it possible to store a small library on a few discs, each the size of a gramophone record, within the next few years. Videodiscs will be viewable via ordinary television screens, and TV may provide the normal basis for finding and using information in the home. It is already possible to track down factual information via a television set using *videotex*. The next step is to employ the system in various other types of transaction. For example, mail order catalogues can be put on television, goods ordered and the money paid from a bank account, all while the viewer is sitting at home.

Just as methods of handling information are becoming more diverse, so, too, are the methods of transmitting it. One of the most spectacular advances in communications during the past two decades has been the growth in the transmission of messages via Earth satellites. *Satellite communication* is now about to make a major impact on the transmission of television programmes, and so on the home viewer. But it is equally significant for the part it can play in the world-wide transmission of very large quantities of *data* of any kind.

Some of the more complex developments in electronic communication still require large (*mainframe*) computers. For example, translation from one language to another (computer translation) is still only moderately advanced, even with powerful machines. Computers that try to duplicate the problem-solving capabilities of human beings (*expert systems*) are likewise at an early (but already useful) stage.

This short account of new information technology is far from comprehensive, but should serve to illustrate its diversity. Almost inevitably, the rapid changes within the field lead to a rapidly changing vocabulary. Confusion results: not only because the new terms may not be easy for all participants to understand, but also because the same term may be used in more than one way. A major purpose of this dictionary is to help dispel such confusion by bringing together and codifying the most important specialized terms currently in use in the various parts of this diverse field.

The diffuseness of the subject matter has made decisions on which items to include and which to exclude particularly difficult. Basic computer technology has been limited to the minimum required for an understanding of most information technology. Terms with a restricted application (such as most manufacturers' brand names) have generally been excluded. Since the emphasis is on information, techniques which are mainly used for entertainment receive less emphasis. For example, 'videotape' is treated for the purposes of this dictionary as less important than 'videodisc'. The main criterion has been that the term should be likely to have a reasonably wide usage for a reasonable period of time.

So far as our intended audience is concerned, it would be true to say that this dictionary is intended for the non-specialist, but it would not be very helpful. Few people would consider the whole of new information technology as their specialism. Rather there are computer specialists whose acquaintance with information terminology may be limited, librarians who may be uncertain of the meanings of some computer-related terms, publishers who may need to learn the jargon of new methods of information handling, and so on. Hence, the contents of this dictionary not only are aimed at non-specialists (that is, at readers with an interest in the field, but with no expertise in any of its branches); they are also designed to aid the various specialist groups whose concerns overlap in the field of new information technology.

Our selection of the words included in this dictionary stems mainly from our own acquisition and vending of periodicals and advertising literature in the field. We hope that, if nothing more, it will help the unfortunate reader through some of the flood of jargon these contain. To give some coherence to the field, a few topics are treated at greater length than the remainder. Such longer entries are intended to act as foci for particular parts of the field: cross-references from and to them allow more specialized entries to be placed in their appropriate context.

Equally, a reader who works through the longer entries should gain a good overall picture of the present state of the art. (If you wish to use the dictionary for learning in this way, the appropriate longer entries are listed after this introduction.) Graphics have been introduced whenever it seems necessary to enhance the verbal descriptions.

A few words should be said in conclusion about the use of this dictionary. An italicized word in any entry means that there is a cross-reference to that word: it can therefore be looked up if you feel unsure as to its meaning. Some words, eg 'computer', have entries in the dictionary, but occur so frequently that they are only italicized in special circumstances. Many words can occur in different forms, eg as a noun or a verb. Only one form is normally given in this dictionary; so it is advisable to check under different headings. Several terms have variant meanings: these are distinguished in their respective entries by separate numbers. If there is likely to be any ambiguity of meaning, cross-references include the relevant entry number. As a general rule, acronyms have been printed in block capitals.

My colleagues and I would like to express our thanks to Arthur Phillips, Dave Adams and Kate Waters for their help in compiling this dictionary. We would greatly appreciate comments from users of this dictionary on any problems they encounter.

Jack Meadows *University of Leicester*

Longer entries providing an introduction to new information technology

artificial intelligence
cable television
computer graphics
computer-aided phototypesetting
document delivery system
electronic journal
electronic mail
expert systems
facsimile transmission
information retrieval system

information retrieval techniques
machine translation
machine-aided translation
satellite communications
speech recognition
speech synthesis
teleconferencing
video disc
videotex
word processing

A and I abstracting and indexing.

AB automated bibliography.

ABC 1. American Broadcasting Corporation. 2. Australian Broadcasting Corporation.

ABCA American Business Communication Association.

ABEND an abnormal end to a computer task due to an error, or an intervention by the operator.

ABES Association for Broadcast Engineering Standards, US.

ABI/INFORM Abstracted Business Information/Information Needs. A *bibliographic database* covering business management and administration (see *INFORM*).

abort to abandon an activity, usually because an error has been made.

abort timer a device which terminates *dial up* data transmission if no data are sent within a predetermined time.

absolute coding *program instruction* in *machine code*.

ABSTI Advisory Board on Scientific and Technical Information, Canada.

abstract an abbreviated representation of the contents of a document. The two most important types of abstract are: a. indicative abstracts. These indicate the content of the document (ie what it is about), rather than its methods and findings; b. informative abstracts. These emphasize the main findings, conclusions, and, if appropriate, methods. Abstracts provided by abstracts journals and information services will typically be a mixture of informative and indicative. The word is also used as a verb to indicate the activity of abstracting.

AC 1. alternating current. 2. automatic computer. 3. analog(ue) computer.

ACARD *Advisory Council for Applied Research and Development.*

ACCC Ad Hoc Committee for Competitive Communications, US.

accent mark placed above or below a character; usually to indicate its pronunciation. (See also *diacritic*.)

access 1. used either as a verb or noun to indicate either gaining control of a system or the acquisition of data from a *storage device* or *peripheral unit*. 2. a US *teleordering* system.

ACCESS Automated Catalog of Computer Equipment and Software Systems, US Army.

access time the time taken to retrieve information from a storage device. For examples of typical access times, see *storage devices*.

Accounts Index *database* compiled by the American Institute of Certified Public Accountants, covering accounting, auditing, banking, finance, investment and related areas. It is available for *on-line searching* via *SDC*, and offers *off-line* services.

accuracy freedom from error, or the size of an error: used in relation to *programs*, *data* and machine operations. Should not be confused with *precision*.

ACIA asynchronous communications interface adapter. A device which *formats* and controls *data* at an asynchronous communications interface.

ACK affirmative acknowledgement sent down a *transmission line* to indicate either that a block of *data* has been received or that the receiver is ready to receive data (see *NAK*).

ACL Audit Command Language: a *high level programming language*.

ACLS American Council of Learned Societies.

ACM Association for Computing Machinery. A US based international organization aimed at advancing computer technology and its applications.

ACOMPLIS A Computerised London Information Service. An information service operated by the Library of the Greater London Council.

acoustic coupler a device capable of transmitting and receiving specified sound tones along telephone lines. It allows a *computer* and *terminal* to be connected via these lines, using a *modem* and telephone handset.

acoustic delay line a *delay line* whose action is based on the time of propagation of sound waves.

ACRL Association of College and Research Libraries. A US organization within the *ALA*.

action frame see *response frame*.

active file a computer *file* in current use.

ACTSU Association of Computer Time Sharing Users, US.

ACU *automatic calling unit*.

ADA a *high level programming language*, adopted by the US Department of Defense as a standard for military systems; but it is also used in other systems.

ADAM automatic document abstracting method (see *automatic abstracting*).

ADAPSO Association of Data Processing Service Organizations, US and Canada.

ADB a Danish *teleordering* system.

ADC *analog(ue)* to *digital* conversion.

added entry in cataloguing, a secondary entry (ie any entry other than the main entry).

address 1. in telecommunications, this refers to the coded representation either of the destination of data, or of the terminal from which the data originate. 2. in computers, it is a number which identifies a location in the computer's memory.

ADI American Documentation Institute.

ADIS Automatic Data Interchange System.

adjacency a term in *character recognition*. It refers to print where the reference lines between two consecutive *characters* are separated by less than a specified distance.

Administrative Support System a *word processing* system aimed especially at business executives.

Adonis an electronic system being developed by a consortium of major European scientific publishers for the storage and supply of full-text documents. It is to be operated in conjunction with existing *document delivery systems*. Journal articles are to be stored on *digital videodiscs* in a form accessible for *on-demand retrieval* in response to inter-library loan requests. In the future, electronic transmission of documents to the US via a satellite link is envisaged.

ADP 1. advanced data processing (see *data processing*). 2. automatic data processing (see *data processing*).

ADPE automatic *data processing* equipment.

ADPS automatic *data processing* system.

ADRES Army Data Retrieval System, US.

ADRS Automatic Document Request Service. A service provided by *Blaise* (a *host* information service). It allows subscribers at an *on-line terminal* to request loans or photocopies of documents from the *British Library Lending Division*.

Advisory Council for Applied Research and Development ACARD is a UK body which advises government and publishes reports on R&D policy (including new information technology).

ADX an *automatic exchange* in a data transmission *network*.

AEBIG Aslib Economics and Business

Information Group (see *Aslib*).

AECT Association for Educational Communications and Technology, US.

AEDS Association for Educational Data Systems, US.

aerial in a radio communication system, this is the device which radiates the transmitted electrical signal into space. Equally, it is the device which receives the signal and feeds it in electrical form into the receivers.

AEWIS Army Electronic Warfare Information System, US.

AFIPS American Federation of Information Processing Societies.

AFR automatic field/format recognition: a computer *input* facility.

Agate a 5½ *point typeface* often used in setting classified advertisements in the US.

AGLINET Agricultural Libraries Information Network (under the aegis of the United Nations).

Agricola *database* produced by the US Department of Agriculture, covering agriculture and related topics. Available via *BRS*, *Lockheed* and *SDC*.

Agricultural Information System (AGRIS) an international agricultural database organized under the aegis of the Food and Agricultural Organization (FAO).

AGRIS *Agricultural Information System*.

AI *artificial intelligence*.

AIDS 1. Aerospace Intelligence Data System of *IBM*, US. 2. automated information dissemination system.

AIRS Automatic Image Retrieval System.

AKWIC Author and Key Word In Context: a form of computerized index (see *KWIC*).

ALA American Library Association.

ALA/ISAD the *ALA*'s Information Science and Automation Division.

ALA print train a standard set of characters drawn up by the American Library Association for use in *machine-readable* bibliographic records.

ALAS Automated Literature Alerting System: a *current awareness service* offered within the context of an *information retrieval system*.

Albert name given by *British Telecom* to a machine intended to combine *teletext*, *telex*, *word processing* and *telephone* functions.

ALGOL Algorithmic Orientated Language. A *high level programming language* used especially for scientific applications (see *algorithm*).

algorithm a procedure, or rule, for the solution of a problem in a finite number of steps. In computing it normally refers to a set of simple rules for the solution of a mathematically expressed problem, or for evaluating a function.

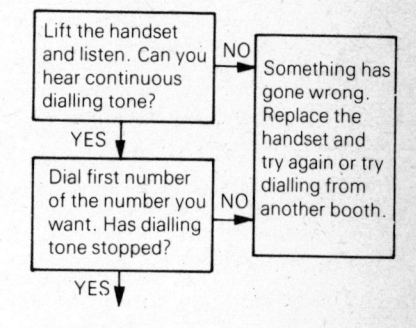

Simple algorithm: using a telephone. The great advantage of this kind of problem-solving system, for computing purposes, is that it reduces the problem to a series of yes/no options, which are easily translated into binary form.

aliasing the removal of the jagged line, or 'step edge' effect, on *graphic displays* (see *computer graphics*).

ALIS a general abbreviation for automated library information system.

ALP automated language processing (see *data processing*, *word processing* and *machine translation*).

ALPAC National Academy of Sciences' Automated Language Processing Advisory Committee, US.

alphabet length the length of a lower-case alphabet in *points*. Used to compare different designs of printers' *typefaces*.

alphageometrics a method for generating *videotex* images on a screen. Displays are constructed out of geometric elements, such as diagonal lines, arcs and circles (see, in contrast, *alphamosaics*).

alphameric synonymous with *alphanumeric*.

alphamosaic a method for generating *videotex* images on a screen. Displays are constructed using a mosaic of dots. Alphamosaics have been chosen in preference to *alpha-geometrics* as a European standard for the next *generation* of videotex systems.

alphanumeric an acronym formed from the words 'alphabetic' and 'numeric'. It signifies that data may contain both alphabetical and numerical information.

alternating current (AC) an electrical current, the direction of which is periodically reversed. The frequency of reversal is usually of the order of many cycles per second.

ALU *arithmetic and logic unit*.

AM *amplitude modulation*.

AMACUS Automated Microfilm Aperture Card Update System. A system which uses an *aperture card* as the primary unit of information *storage*.

America: History of Life a *bibliographic database* covering US and Canadian history and affairs, available via *Lockheed*.

AMFIS Automatic Microfilm Information System (see *COM*).

amplitude the peak positive or negative value of a wave (or signal).

amplitude modulation a form of *modulation* in which the *amplitude* of a *carrier wave* is varied by an amount proportional to the amplitude of the modulating signal.

AMR *automatic message routing*.

AMTD Automatic *Magnetic Tape* Dissemination service offered by the US Defense Documentation Center.

analog(ue) representation of information by an output signal which varies in a continuous manner with respect to the input. It is to be contrasted with *digital* representation of information.

AND an AND *gate* is used in computer *logic* to combine *binary* signals in such a way that there is an output signal only if all input channels carry a signal. For the case of two input signals, this leads to the following table.

Input 1	Input 2	Output
1	1	1
1	0	0
0	1	0
0	0	0

AND gate a *gate* that implements the logic of the *AND* function.

angle modulation see *modulation*.

ANIK a series of Canadian *communication satellites* launched from the early 1970s onwards. (ANIK is 'brother' in Eskimo).

ANSI American National Standards Institute. It has established many standards in the fields of computing and information handling which are accepted world-wide.

answerback see *voice answerback*.

answerphone a device for automatically responding to telephone calls, recording any messages for playing back later.

Antiope a French *videotex* (*viewdata* and *teletext*) system. The name is an acronym for 'L'Acquisition Numérique et Télévisualisation d'Images Organisées en Pages d'Ecriture'. (The Numerical Acquisition and Televisual Display of Images organized into Pages of Text.)

Antiope-Didon the *teletext* component of the French *Antiope videotex* system.

Antiope-Titan *viewdata* component of the French *Antiope videotex* system. Often referred to by the brand name 'Teletel'.

AOIPS Atmospheric and Oceanic Information Processing System. Employs *computer graphics* to enhance images received from meteorological satellites. (See also *DIDS*.)

AP *attached processor.*

aperture card an 80-column card (*punched card*, or *edge-punched* card) which has a 35 x 48mm frame of microfilm inserted. Aperture cards can be used to form an index, which is sorted via the punched holes. After sorting, more extensive information is immediately available on the inserted microfilm.

APILIT *database* produced by the American Petroleum Institute, covering petroleum refining, storage and transportation; petrochemicals, petroleum products and petroleum substitutes. Available via *SDC*.

APIPAT as for *APILIT*, but covering patents.

APL A Programming Language; a *high level programming language* suitable for use on *mainframe* computers with large *memories*. It is sometimes used in conjunction with statistical *databases*.

APOLLO Article Procurement with On-Line Local Ordering. An experimental *document delivery system* run by the *CEC* and *ESA*.

application program see *applications package*.

applications package a *program*, or set of programs, designed to perform a particular application, or task (as in *information retrieval*, *word processing*, *data analysis*).

applications software *programs*, or packages (see *applications package*), designed to carry out specific tasks, or applications; as distinct from *systems software*, which controls the operation of the total computer system.

APT Automatically Programmed Tools. A computer language used for control of machine tools.

AQL acceptance quality level. A general engineering term, usually used to refer to the performance of machines and components: in particular, their breakdown or failure rate.

architecture the way in which computer *hardware* and *software* interact so as to provide the type of facilities and performance required.

archival storage see *backing storage*.

archive used either as a verb or noun to indicate: a. the process of storing data *files* in a retrieval form; b. the data files so stored.

archive diskette synonymous with *diskette* or *floppy disc*.

ARDIS Army Research and Development Information System, US Army.

area composition in *phototypesetting*, usually refers to putting together a *page*, either by use of an automatic page *make-up program*, or by an operator calling up the necessary material for display at a *VDT*.

area search a term used in information retrieval for a search of those items within a database which make up a single group, or category (see *information retrieval system*).

arithmetic and logic unit (ALU) the *microprocessor* in the *central processing unit* which executes the arithmetic and logical operations required by an *input* command.

ARPANET a *resource-sharing computer network* supported by the Advanced Research Projects Agency of the US Department of Defense.

ARQ automatic request for correction. A system which provides error correction by requesting retransmission of mutilated characters. (See also *automatic request for repetition*.)

Art Bibliographies Modern a *database* covering literature on art from the beginning of the 19th Century. Accessible via *Lockheed*.

ARTEMIS an acronym for Automatic Retrieval of Text through European Multipurpose Information Services. It is planned to be a *document delivery system* which will supply journal articles *on demand* via *facsimile* transmission or *digitized* text. Delivery will be either by a *facsimile receiver*, or as print out from a *teleprinter*. A fully integrated ARTEMIS system will require the establishment of a set of standards between *hosts*, *databases* and users. The European Commission is currently working towards this objective.

artificial cognition the ability of a machine to sense a *character* by optical means, and then to determine its nature (by comparing it with a set of standard characters).

artificial intelligence (AI) artificial intelligence concerns the design of intelligent *computer* systems: that is, systems which exhibit the characteristics commonly associated with human intelligence – understanding *natural language*, problem-solving, learning, logical reasoning, etc.
Computers are well suited to handling those forms of reasoning and problem-solving which can be clearly broken down into a series of 'logical' steps, eg the performance of numerical calculations. However, other aspects of intelligence cannot be so easily programmed: it is the goal of AI to overcome these difficulties.
The nature of the difficulties is often illustrated in terms of computer chess. The computer is programmed to consider the possible moves it could make, and the possible responses from an opponent. It then evaluates the outcome of each sequence of moves in accord with prescribed criteria, and selects the best option. A computer can 'search' thousands of moves in the time a human can only consider a few, but no computer can, as yet, beat a human chess master. The master's advantage appears to lie in a form of intelligence derived from experience, and produces an ability to draw crucial inferences from the pattern of the pieces, and the opponent's pattern of play. The reasoning subsumed within the chess master's intelligence cannot be clearly explicated into steps, and cannot therefore be incorporated into a computer program. Nevertheless, computers can now play chess and other similar types of game to a high standard. These are generally referred to as problem-solving forms of AI. Similar problem-solving techniques, based on the principles of 'search' and 'problem reduction', have a variety of more practical uses, eg in performing mathematical integration of complex equations.
Related to 'problem solving' is 'logical reasoning'. In exploring this field, AI techniques have been used to develop methods for searching information in a *database*, so as to test the validity of generalized statements ('theorems'). The technique has been extended to include monitoring the acceptability of theorems as information is added to the database. Such systems can also identify crucial data, eg those which are anomalous in relation to a specific theorem, so that they can be scrutinized in more detail. A further development is the so-called '*expert system*' – a form of computer-based consultant. It is the AI component which distinguishes the expert system from a specialized *on-line information retrieval system*; allowing a dialogue, rather than a simple interrogation of a database. The understanding of language is a further area of application for AI which is of fundamental importance to information systems. For clearly, if AI could resolve all the inherent problems, computers could receive natural language as *input*, and perform automatic translation into *machine language*, or, indeed, into other natural languages (see *machine translation* and *HAMT*). Somewhat similar in impact to the understanding of language is the field of visual

pattern recognition. If computers could recognize objects (via, for example, television cameras), this would offer even greater flexibility in the input of information to computer systems. AI of this type is crucial in the field of *robotics*.

ARU *audio response unit.*

ASA American Standards Association (formerly the USASA) has groups responsible for the establishment of standards in the field of *data processing*.

ASC automatic sequence control: a *program* feature.

ASCA Automatic Subject Citation Alert. A computer-produced *current awareness* service based on the *database* of the *Science Citation Index.*

ascender a typographic expression indicating that part of a lower-case character which rises above the normal body height, eg the upper part of the letters 'b' and 'h'.

ASCII (code) a US standard computer *code*, adopted in Europe, in which eight binary *bits* can be combined to represent the characters on a typewriter keyboard. Only seven bits (128 possible combinations) are necessary to describe all the characters. The eighth bit is either used for error checking purposes, or it remains unused. Most commercial VDU's and printers utilize the ASCII code.

ASDI automated selective dissemination of information (see *selective dissemination of information*).

ASI American Statistics Index. *Databank* of American social, economic and demographical statistics, searchable via *Lockheed* or *SDC*.

ASIS American Society for Information Science.

Aslib originally stood for Association of Special Libraries and Information Bureaux, but is now used as a name in its own right. Aslib is a British association with headquarters in London.

aspect in information retrieval, this refers to those features of the contents of documents, etc, which are represented by *index term*, *descriptors*, etc.

aspect card a card containing numbers which record the location of *documents* used in an *information retrieval system*.

ASR answer, send and receive. A *teletypewriter* and receiver used in conjunction with a computer.

ASSASSIN Agricultural System for Storage and Subsequent Selection of Information. A bibliographic *information retrieval system* available from ICI.

assemble to put together a *machine language program* from a *symbolic language* program.

assembler a computer *program* which transfers a *symbolic language* program into a *machine language* program. The latter can then be directly executed by the computer.

assembly language a programming language in which each statement corresponds to a single *machine language* instruction. It is normally written in some form of *mnemonic code*.

assigned-term indexing see *indexing*.

assignment indexing a form of *automatic indexing*, less frequently used than *extraction indexing*. Words are selected in a similar fashion as for *extraction indexing*, but instead of using extracted terms as index terms, the extracted terms are used in conjunction with some kind of *thesaurus*, to produce a list of index terms chosen from a *controlled vocabulary* list. In this sense index terms are 'assigned'.

associative storage computer *storage* where locations are identified by their contents, rather than by their names, *addresses*, etc. It is also called 'content-addressed memory' or 'parallel-search storage'. In associative storage, any *keyword* used in the search is compared simultaneously with all the keywords in the store, seeking for a match. More *logic hardware* is required than with conventional storage.

asynchronous transmission a telecommunications term. Transmission of data where time intervals between transmitted characters may be of unequal length. Asynchronous communication with other devices does not require a continuous exchange of synchronization signals. (See also *synchronous transmission*.)

AT&T the American Telephone and Telegraph Company. Claimed to be the world's largest corporation. It is the largest common carrier in the US, handling over 80 per cent of the installed telephones in the US (although serving about a third of the geographical area). Runs the 'Bell System' and Western Electric, and jointly runs the Bell Telephone Laboratories, one of the world's largest research organizations.

ATLAS Automatic Tabulating, Listing and Sorting System. A *software package* extensively used for the purposes indicated.

ATM see *automatic teller machine*.

ATS Application Technology Satellite. The name attached to a series of US communications satellites. Both audio and video channels are used, particularly for medical communication (see *satellite communication*).

attached processor a *processor* attached to a *central processor*, often sharing its memory.

attenuation in communications systems, this refers to the loss in signal strength encountered over long transmission paths.

AUDACIOUS Automatic Direct Access to Information with On-line UDC System. An *interactive* retrieval system using *UDC* for the *coding* and search scheme (see *information retrieval systems* and *on-line searching*).

audio capable of being heard by the human ear (ie within the frequency range 15 Hz (cycles/sec) to 20,000 Hz).

audio cassette a small cartridge containing

magnetic recording tape mounted on rotatable reels. Size has been standardized to allow interchange between a wide variety of devices.

audio response unit a device which can connect a telephone to a computer in order to provide voice response to a user's enquiries.

audit trail a trail followed through a *data processing* system to check that the steps involved lead back from the computer *output* to the original document.

AUSINET an Australian *network* offering nation-wide on-line access to a range of *databases* (see *on-line searching*, *CSIRONET* and *MIDAS*).

authority file a set of records identifying a standard for established forms of headings, *index terms*, or other items, which may be subsequently used for information retrieval. An authority file may also contain established cross-references. A *thesaurus* is one example of such a file.

auto abstract also called an *automatic abstract*, it is an '*abstract*' produced by a computer analysis of a document (see *automatic abstracting*).

autoanswer a device which automatically answers calls via the telephone network (see also *auto dialler*).

autocall see *automatic calling unit*.

autocode a system for the computer conversion of *symbolic code* into *machine code*.

autodial see *auto dialler*.

auto dialler a device which permits automatic dialling of calls via the telephone network. (See also *automatic calling unit*.)

auto-kerning in *phototypesetting*, the automatic reduction of the spacing between certain characters when they appear together, eg 'A' and 'V' in 'AV'.

automated bibliography a bibliography stored in a computer *file*.

automated dictionary a form of *automated lexicon* used in *machine-aided translation* systems. In contrast with *automated glossaries* it separately lists roots, eg inform and affixes, eg mis-, -ed, -er, -ing, etc. Used by the *LEXIS* and *TEAM* systems (see *machine-aided translation*).

automated glossary a form of *automated lexicon* used in *machine-aided translation* systems. In contrast with *automated dictionaries*, it contains whole words and thus presents many variations of a generic root, eg informed, informer, informing. Used by the *SMART* translation system (see *machine-aided translation*).

automated lexicon a generic term covering all forms of *automated dictionary* (single-word entries) and *automated glossary* (multiple-word term entries, eg 'electron spin resonance'). Automated lexicons constitute the central component within a machine-aided translation system (see *machine-aided translation* and *terminology bank*).

automated stock control use of software on a computer to check on receipt and delivery of goods, including keeping accounts and forecasting demand.

automated thesaurus a computer-based *thesaurus* used in conjunction with an *automated lexicon* within a *machine-aided translation* system.

automatic abstract also called an *auto abstract*; it is an *abstract* produced by a computer analysis of a document (see *automatic abstracting*).

automatic abstracting high-frequency substantive words in a document are identified using the same techniques applied in *automatic indexing* (see *extraction indexing*). Sentences which are found to contain the highest concentration of high-frequency words are then identified and printed out in sequence. The product, an *auto abstract* or *automatic abstract* does not necessarily look like a normal human-prepared abstract, but it normally gives a fairly good indication of what a document is about.

automatic calling unit a device which permits automatic dialling of calls via the telephone network: normally used in business information systems. (See also *auto-dialler*.)

automatic paper carriage a device for guiding, or holding, prior to printing. It feeds sheets, or continuous paper, to the writing heads.

automatic dictionary a component within a *computer translation* system which provides word-for-word substitution from one language to another.

automatic indexing the *automatic* production of an index for documents in a *database*. The most commonly used techniques are described under *extraction indexing* (see also *assignment indexing* and *automatic abstracting*).

automatic logon (log on) a facility offered by *intelligent terminals* when used for *on-line searching* (see, eg *Userkit*). The various *passwords*, identifiers and *addresses* necessary to use local or international tele-communications *networks* and *host* computers are stored at the terminal, so that the *log on* process can be carried out with one, or a few, *key strokes*.

automatic message routing automatic directing of incoming messages to one or more outgoing circuits according to the content of the message.

automatic speech processing this term covers a number of different processes, such as *digital transmission* and storage of voice messages, and *speech recognition*.

automatic stop a means of stopping a computer operation when an error is detected, by an automatic checking procedure.

automatic teller machine a device for providing an automated banking service: eg cash dispensers, balance reports (see *bank on-line teller system*).

automation process (or result) of making machines self-acting and/or self-controlling, by eliminating the need for human intervention in the process.

autoplotter a *plotter* which produces graphical output under *computer* control.

autoscore the automatic underlining of text.

auxiliary storage computer *storage* which is external to the computer itself, eg *magnetic disc*, *magnetic tape*.

availability when a computer has its power switched on, its 'availability' is the percentage of time it is available to the user.

AVIP see *BAVIP*.

AVLINE Audiovisual On-line. A *bibliographic database* containing citations to, and abstracts of, audiovisual teaching packages in the health sciences.

babble interference between two or more data transmission channels.

background in *time-sharing* computers, this refers to low-priority tasks which the computer only carries out when not occupied with high-priority items (the *foreground*). Thus background tasks may be carried out while data are being input for a foreground task.

backing storage storage within a computer's main memory is both expensive and limited in amount. When the computer needs access to large quantities of data, additional storage capacity is therefore required. At present this 'backing storage' is normally held on magnetic devices, eg *tapes*, *discs*, *card files*, and *drums*. While these have cost advantages over main 'core' memory, *access times* are slower.

back number any issue of a periodical which precedes the current issue.

back-up a procedure or facility which allows a user to retain information in the event of a computer failure.

back-up (equipment) equipment available for use as a substitute in the event of failure of the equipment normally used. Purchase of such back-up equipment is becoming increasingly common.

backward recovery recovering from a system failure by reversing processes that have already been applied, thus reconstituting the *file* to an earlier state.

BAM see *Basic Access Method*.

band 1. a range of frequencies. 2. Recording area on a magnetic tape or drum.

band-pass the range of frequencies of signals passed by a *filter* without significant attenuation.

bandwidth a telecommunications term. It describes the extent of the frequency spectrum within which signals can pass through a system without significant *attenuation*. Signals carrying information, eg the human voice, are made up of a number of component frequencies. For faithful transmission and reception of a signal, the bandwidth must be sufficient to pass these component frequencies. Unwanted signals, eg *noise*, can sometimes be suppressed by the use of filters of specified bandwidth (see *band-pass*).

bank on-line teller system a system of *automatic teller* machines linked to a central *computer* in *conversational mode* (see *debit card systems*.)

bar-code a type of code used on labels to be read by a *wand* or *bar-code scanner*. The main application is in labelling retail products, the *wand* being used to record the sale at the place and time of purchase, but it is also used to label documents in libraries.

Typical example of a bar-code: the universal product code, perhaps the most widely-seen. The detailed storage of information is shown. For an example of the practical use of the code, see optical character recognition.

bar-code scanner an optical device which can read *data* from documents bearing characters recorded in the form of parallel bars. The characters are translated into *digital* signals for storage or processing (see *bar-code*).

bar printer a *printer* with *character* heads mounted on a type bar.

barrel printer see *drum printer*.

baseband a frequency *band* used for the transmission of picture and synchronization signals in television and some telephone systems.

BASIC Beginner's All-purpose Symbolic Instruction Code. A *high level programming language* designed for ease of use. It is particularly suitable for entering and running *programs on-line*. It is now a standard programming language, in a number of variant forms, for *microcomputers*.

Basic Access Method a computer access method in which each *input/output statement* causes a machine input/output operation to be performed.

Batab a US *teleordering* system.

batch a collection of computer *transactions* processed as a single unit.

batch processing a method of processing data by computer which accumulates *transactions* and processes them as a single unit (a *batch*), rather than as they arise.

baud a unit of signalling speed in telecommunications. The speed in bauds is the number of discrete signal events per second. If each signal event represents one *bit*, a baud corresponds to bits per second, otherwise the baud rate cannot immediately be equated to bits per second.

Baudot code standard five-*channel telex* code. The name derives from Emil Baudot, who produced the first major teleprinter code in the nineteenth century. In the UK, the telex code is called the Murray code.

BAVIP British Association of Viewdata Information Providers (see *viewdata* and *information provider*).

BBC British Broadcasting Corporation. The state-owned radio and television broadcasting agency in the UK.

BC see *Bliss classification*.

BCD *binary coded decimal*.

BCPA British Copyright Protection Association.

BCS British Computer Society.

bead a small *program* for a specific function. (See also *thread*.)

beam store a *storage* device in which electron beams are used to write, or read, data.

beamwidth the angular extent over which an antenna can readily detect or transmit a signal.

BEEF Business and Engineering Enriched FORTRAN. An adaptation of the *FORTRAN programming language* to enhance its capabilities for business and engineering applications.

beginning of file label a *record* at the beginning of a *file* which gives information about the file's contents and limits.

beginning of information marker (BIM) usually an area of reflecting material on a *magnetic tape* which indicates the beginning of the area on which information is recorded. Some systems use transparent, instead of reflecting, material, or some kind of perforation in the tape (see *magnetic tape*).

Bell and Howell Newspaper Index a very large *database* (over 2,000,000 entries) covering news and current affairs items appearing in major US newspapers from 1977 onwards. Accessible via *SDC*.

bell character the member of a *character* set used to sound a bell on a *terminal* device. Also used as a *flag* in some *phototypesetting* systems (see *Bell code*).

Bell code a computer *code* used for *phototypesetting* commands.

Bell System the network of telephone and data circuits, *switching* offices, television and other links in the US operated by *AT&T*, its subsidiaries and associated companies.

Beltel proposed *viewdata (interactive videotex)* system for South Africa.

benchmark a program designed to test

and compare the performance of different computers.

BESSY Bestell-system. A German *teleordering* system.

Betamax former name of *Beta*.

Beta (system) a type of *video cassette recording* system, developed by Sony. Uses slightly smaller cassettes than the *VHS* system.

BEX *broadband* exchange.

bibliographic(al) relating to the description of documents.

bibliographic coupling a method of grouping documents by examining the number of common *citations* that the documents make. Thus, if two documents each cite three other documents in common, the citing documents are said to have a bibliographic coupling strength of three (see *co-citation indexing*, *information retrieval techniques*).

bibliographic database a *database* containing information relating to documents (books, articles, reports, etc). The information normally covers details of authorship and title, together with place and date of publication. Some databases include *abstracts*. The content of each item is indexed to facilitate search and access (see *information retrieval systems*).

biconditional synonymous with *equivalence*.

BIDAP Bibliographic Data Processing Program. A *software package* for the *processing* of bibliographic data.

bidirectional printing the direction of printing of consecutive lines is alternated so as to eliminate the need for a return to the beginning of a line. This speeds up the printing process.

Bildschirmtext a West German *viewdata* (interactive *videotex*) system.

billi- prefix meaning one thousand million

(10^9). The US billion ($= 10^9$) is coming into use in the UK, but the UK billion ($= 10^{12}$) is still used.

BIM *beginning of information marker*.

binary a system in which the choice is limited to one of two alternatives, eg an on-off switch.

binary chop a method of computer searching a stored table for an item by successively splitting it in half and searching one half for the presence of the item. If found, the half is split in two once more, and the process repeated. This continues until the exact position of the item is located.

binary coded decimal each individual decimal digit is represented by the corresponding group of binary digits. The resulting number is not the same as a number in *binary notation*. For example, the number 15 in binary coded decimal is 0001 0101, whereas in binary notation it is 1111.

binary digit a digit on the binary scale of notation; either 1 or 0.

binary notation the writing of numbers to the base 2, so that the position of the digits in a number designates powers of 2. For example, the number 101 represents $1 \times 2^2 + 0 \times 2 + 1 \times 1$ ($= 5$ in decimal notation). The table of binary numbers from 0 to 10 is as follows:

decimal notation	binary notation	decimal notation	binary notation
0	0000	6	0110
1	0001	7	0111
2	0010	8	1000
3	0011	9	1001
4	0100	10	1010
5	0101	11	1011

binary number any number written in binary notation.

binary search a method of searching for an element in a table or *serial file*; the location is successively narrowed down by halving the table or file. (See also *binary chop*.)

bind the action of transforming two, or

more, *object program* modules into a program for execution.

biosensor a device for detecting and transmitting data relating to biological activity, so that they can be processed, displayed and stored.

Biosis Previews a large *database* containing approximately 2,500,000 items. The subject range is Biology and the Life Sciences as covered by *Biological Abstracts* and the *Bio Research Index*. It is accessible via most of the major *hosts*.

BIS *British Imperial System*.

bisynchronous the continuous exchange of *synchronization* signals between communications devices.

bit an abbreviation of binary digit. It represents the smallest unit of information (corresponding to, eg 0 or 1; 'on' or 'off'; 'signal' or 'no signal'). Computers usually store information as a series of bits. (See also *byte*.)

bit density the number of bits contained in a storage area, eg of a *magnetic tape* or *magnetic disc*.

bite alternative spelling of *byte*.

bit location a *storage* position in a *record* capable of storing one *bit*.

bit parallel see *parallel bit transmission*.

bit position refers to a position in a bit sequence. For example, to represent 2 in binary form requires a 1-bit in the second position (10); to represent 4 in binary form required a 1-bit in the third position (100), etc.

bit rate the speed at which *bits* are transmitted, usually expressed in bits per second. (See also *baud*.)

bit serial the sequential transmission of the bits in a group through a single channel.

bit-slice microprocessor a *microprocessor*

formed from other microprocessors of shorter *word* length.

bit-slice processing *microprocessors* which allow large scale parallel *data processing* (ie permit many *jobs* to be done simultaneously).

bit stream a set of related *bits* (a *bit string*) travelling along a communication line.

bit string a set of related *bits*.

black box a device that performs a specific function, but whose detailed operation is not known, or not specified, in the context of the discussion.

Blaise (BLAISE) British Library Automated Information Service. Contains bibliographical details of all books published in the UK since 1950, and in the US since 1968. In addition, it acts as a *host* and it provides access to a number of medical and chemical *databases*, eg *Medline*, *Chemline* and *Toxline*. Blaise additionally offers a cataloguing service. Specially developed *software* enables Blaise's *Marc* records to be retrieved and edited to produce local catalogues.

bleed spreading of ink beyond the edges of a printed character: a problem in *optical character recognition*.

blind 1. a piece of equipment which is unreceptive to data. 2. to make a piece of equipment unreceptive.

blind keyboard a *keyboard* which does not provide a visual display, or *hardcopy* of data entered through the keyboard.

blinking a method of signalling important messages by flashing characters on a display screen.

blip 1. an unwanted signal on a display screen. 2. a document mark.

blip counting a position-sensing technique based on adding or subtracting one from a *location register*, depending on the direction in which each position mark (blip) passes a sensor.

Bliss classification a classification scheme which uses 26 alphabetic classes (A-Z), with subdivisions of each main class indicated by the addition of further letters.

BLLD *British Library Lending Division.*

block 1. a group of information units handled as a single unit. In many *magnetic storage* devices only complete blocks can be accessed, or transferred. Block size may be fixed or variable, depending on the equipment. 2. a *half-tone* or *line printing* plate.

block diagram a diagram of a system or a computer *program* in which the parts are represented by boxes, usually labelled with interconnecting lines. (See also *flow diagram*.)

blocking the creation of *blocks* from individual *records*.

block transfer the movement of *data* in *blocks*.

blow back 1. full size print-on-paper copy of information stored on *microform*. 2. image enlargement on a *cathode ray tube*.

blowing as in '*PROM* blowing', this refers to *programming* read only memory *(ROM)* using special equipment.

blow up an enlargement, usually of pictorial matter.

body (size) a term used in *typesetting* to describe the size of typeface in *points*. 'Body' is also used to describe the viscosity of printing ink.

bold face a typeface which appears blacker than normal. Usually employed to give emphasis, eg to headings.

bomb loss of a computer *program* due to incorrect commands.

book in the computing sense, a large segment of computer memory.

book message a message which is sent to two or more destinations.

Bookseller Data a Danish *teleordering* system.

Boolean algebra an algebra dealing with classes, propositions, etc, associated with such operators as AND, OR, NOT, IF, THEN, EXCEPT, etc: it contrasts with conventional algebra, which deals with mathematical relationships. Though developed in the 19th Century, Boolean algebra has wide applicability in *computer* analysis of information and problems, because it expresses logical relationships in a form which can be accommodated within the *binary logic* of *digital computers*. This type of logical expression and analysis is usually illustrated by means of a *truth table*. The example below asks 'if X may be true or false, and Y may similarly be true or false, which combinations of truth and falsehood of X and Y lead to a true or false Z?' In the example given, '0' corresponds to 'false' and '1' corresponds to 'true'. It can be seen that the Z postulated in this relationship is true if one of X or Y is true, but not if both X and Y are true or false. Complex logical relationships can be broken down into binary elements in this way. (The name derives from George Boole, a British mathematician, who first developed this type of algebra.)

X	Y	Z
0	0	0
1	0	1
0	1	1
1	1	0

Boolean calculus synonym for *Boolean algebra.*

Boolean logic synonym for *Boolean algebra.*

Boolean operation an operation in which the result of giving each of a number of operands one of two possible values, is itself one of two values. Such operations can clearly be expressed in a *binary* form for computer analysis (see *Boolean algebra*).

Boolean operation table a table in which

a *Boolean operation* is expressed: for each combination of the one of two values of each operand, the one of two possible values of the result is shown. When the two possible values for each element are 'true' or 'false', the table is known as a *truth table* (see *Boolean algebra*).

boot short for *bootstrap*.

booting short for bootstrapping, this term is usually used to refer to the transfer of a *disc operating system program* from its *storage* on a *disc* to a *computer's* working *memory*. In computer jargon, an operator may '*boot the disc*', '*boot the DOS*', or '*boot DOS*'.

bootstrap a method of inputting data prior to the *loading* of a computer *program*, so causing the program to be loaded.

Boris a Canadian *viewdata* (interactive *videotex*) system.

BOT beginning of tape (see *magnetic tape*).

BPCC British Printing and Communication Corporation (formerly British Printing Corporation).

BPI bits per inch: used for measuring density of data on a storage medium. (See also *bits*.)

BPS bits per second. (See also *bits*.)

branch a point in a *program* where a computer must select one out of two, or more, pathways.

breadboard an experimental set up of an electronic circuit for design and operational testing.

brightness ratio a measure of *contrast*. It refers to the ratio between the brightest and darkest parts of a printed paper sheet. The term is used in the context of *optical character recognition* and *facsimile transmission*.

Brisch classification a *classification and coding system* covering every facet of the activities of engineering organizations (see *classification and coding systems*).

British Imperial System system of units of measurement from which the *US Customary System* was developed. It uses such units as feet and inches, pounds and ounces, and pints and gallons. It is being superseded by the *SI* system in most scientific and technical areas.

British Library Lending Division the major UK supplier of documents on inter-library loan.

British Telecom British Telecommunications. The UK Post Office has been split into two parts: the first (called The Post Office) dealing with conventional letter and parcel post, and the second (British Telecom) dealing with telecommunications services (*telephone*, *telex*, *telegraph*, *Datel*, *Prestel*, etc).

broadband communication channel with a *bandwidth* greater than a voice-grade channel, and therefore capable of higher-speed data transmission.

Broadband Exchange public *switched* telecommunications *network* of the *Western Union*, US.

broadcast the dissemination of information to several receivers simultaneously (usually via electromagnetic signals).

broadcast satellite a form of frequency allocation for *communications satellites* which identifies the *uplink* stations only.

Broadcast Satellite Experiment a Japanese *communications satellite* particularly intended to investigate direct television transmission. (See also *Experimental Communications Satellite*, *direct transmission satellite*.)

BROWSER Browsing On-line with Selective Retrieval. A system which offers automatic *natural language searching* of documents in a *database*. All documents entering the database are subjected to a form of automatic *extraction indexing* which results in their characterization by *strings* of words, weighted in accord with their frequency of occurrence.
An English language search query can then

be accommodated, by comparing terms used in the query to those generated and weighted by the *automatic index*. Documents are selected on the basis of 'best fit', retrieved and listed in ranked order (see *LEADERMART* and *SMART*).

BRS Bibliographic Retrieval Services. A *host*, based in the US, offering access to about 25 of the most widely used *bibliographic databases*.

brush an electrical device for reading *information* from a *punched card*.

BSE *Broadcast Satellite Experiment*.

BSI British Standards Institution: establishes standards for measurement, nomenclature and product performance in the UK.

BTAM 1. Basic Terminal Access Method: *Basic Access Method* from a *terminal*. 2. Basic Telecommunications Access Method: *Basic Access Method* using telecommunications channels.

BTX an abbreviation commonly used for *Bildschirmtext*, a West German viewdata (interactive *videotex*) system.

bubble memory see *magnetic bubble memory*.

bucket in computing, a place, or unit, of storage.

buffer 1. a *storage* device (typically between *input/output* equipment and the computer) where *information* is assembled to allow for differences, eg in data flow rate, in its onward transmission. 2. a *circuit* used to isolate one circuit from another.

buffer channel a method for interfacing devices with a computer, which contains *memory addressing* capability and the ability to transfer *words*.

bug an error in a computer *program*.

bulk storage large volume storage, for which *access times* are relatively slow (see *backing storage*).

Bundespost the *PTT* of West Germany.

bureau see *computer bureau*.

Bureau of Standards US Government agency concerned with standards for measurement and performance. Should not be confused with *ANSI*.

burst a set of characters grouped together for *data transmission*.

burst modem in satellite communications, each station sends high-speed bursts of data which are interleaved with each other. These bursts have to be very precisely timed, and are therefore sent using a burst *modem*.

burst traffic transmission of data in bursts, rather than continuously. A simple example is the exchange of information in a telephone conversation.

bus an interconnected system path over which information is transferred, from any one of many sources to any one of many destinations, the devices involved being connected in parallel.

bus driver a power amplifier used to drive several devices using a *bus*.

byte a group of adjacent *bits*, such as 4, 6 or 8 bits, operating as a unit. For example, a six-bit byte may be used to specify a letter of the alphabet, and an eight-bit byte may be used to specify an instruction or an *address*. Normally shorter than a *word*. Unless otherwise indicated a byte is normally assumed to be 8 bits long.

© international sign indicating the copyright assignment of a document.

C Coulomb: the *SI* unit of electrical charge.

CAAS Computer Assisted Acquisition System: to assist libraries in their acquisition of material.

CAB Abstracts *database* compiled by the Commonwealth Agricultural Bureaux, UK, covering a broad area of agriculture and related sciences. The database is accessible via *Lockheed* and *ESA-IRS*.

cable one or more conductors contained within a protective sheathing. If multiple conductors are present, they are electrically isolated from each other. (See also *coaxial cable*.)

cable casting refers to dissemination of information via cables, eg *cable TV*, instead of broadcasting, eg radio, or broadcast TV.

cable television cable television is sometimes called *CATV* (which stands for community antenna television). This is because cable television originally described a system where a communal antenna (aerial) received a broadcast television signal. The signal was then transmitted to domestic television sets via *coaxial cable*. Today, in many systems the cable runs directly from the office distributing the program(me)s to the receiving sets, and no antenna is required. The original purpose of CATV was to improve television reception in difficult areas, but it has developed at markedly different rates in different countries, and for a variety of reasons. In some countries, for example, regulatory authorities have limited its growth. In the US, an anti-trust legal decision in 1968 effectively encouraged the development of CATV, and it has grown greatly since. Diagrams overleaf. CATV may help the spread of new information technology. The coaxial cable which carries the television signal has an *information-carrying capacity* roughly a thousand times greater than the normal telephone cable, and can be used for other purposes besides transmitting the television signal. A disadvantage is the expense of lay-

ing cables and the consequent problem of operating economically in sparsely populated regions.

Not all the channels offered by modern cables are used, so other services can be provided. CATV companies already provide relatively inexpensive information services such as weather reports, news headlines, etc, on these channels. Moreover, each channel can be subdivided into further channels of smaller *bandwidth*. These can be used to provide additional information, eg data or *facsimile* transmission.

Interactive use of CATV increased in the US after a 1972 *FCC* ruling that such systems should be constructed with a 'reverse channel' capability. With this, the subscriber can respond to signals arriving at the television set by sending back signals along the cable. These signals can be fed into a computer thereby providing access to a range of new services. A main use so far has been for simple entertainment, eg subscribers can rate performers on talent shows in *real time*. However, many other services – news, financial, *computer-aided learning*, *information retrieval* from *databases* – are under consideration. It may also prove possible to widen the range of facilities by linking separate CATV systems together to form a larger *network*. These various proposed services overlap to some extent with those planned for *viewdata* systems. One of the main differences between CATV and viewdata is that the latter is transmitted along conventional telephone lines which have a lower capacity.

CAD *computer aided design*.

CAD/CAM computer aided design and manufacture (see *computer aided design* and *computer aided manufacture*).

CAI computer assisted instruction (see *computer-aided instruction*).

CAIC computer assisted indexing and classification ((see *automatic indexing*).

CAIP Computer-Assisted Indexing Program. A *CAIC* system developed by the United Nations.

CAIRS Computer-assisted information

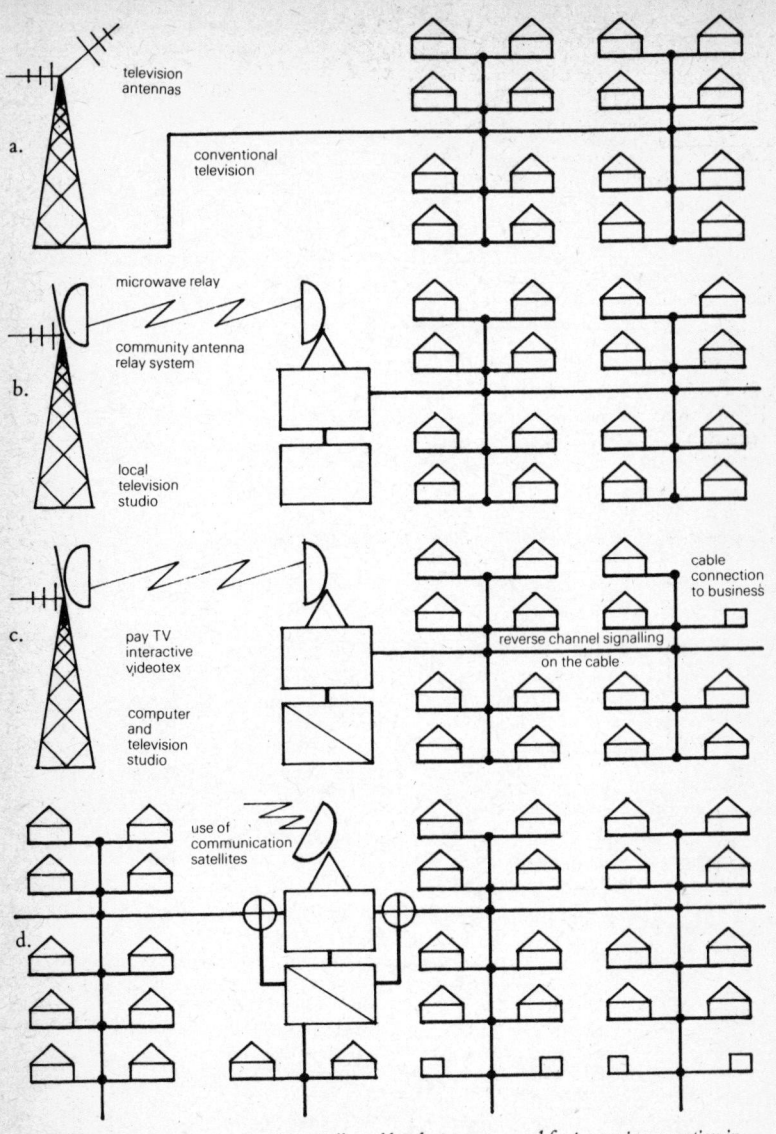

Progressive development of CATV. Initially, cable television was used for improving reception in difficult, eg mountainous, areas (a). It was then developed as an alternative to conventional television systems, with its own television origination (b). The greatest advance was made after the introduction of a reverse channel on the cable, allowing two-way communication (c). Currently, different CATV networks are being interconnected, producing nationwide coverage by CATV stations, aided by communication satellites (d).

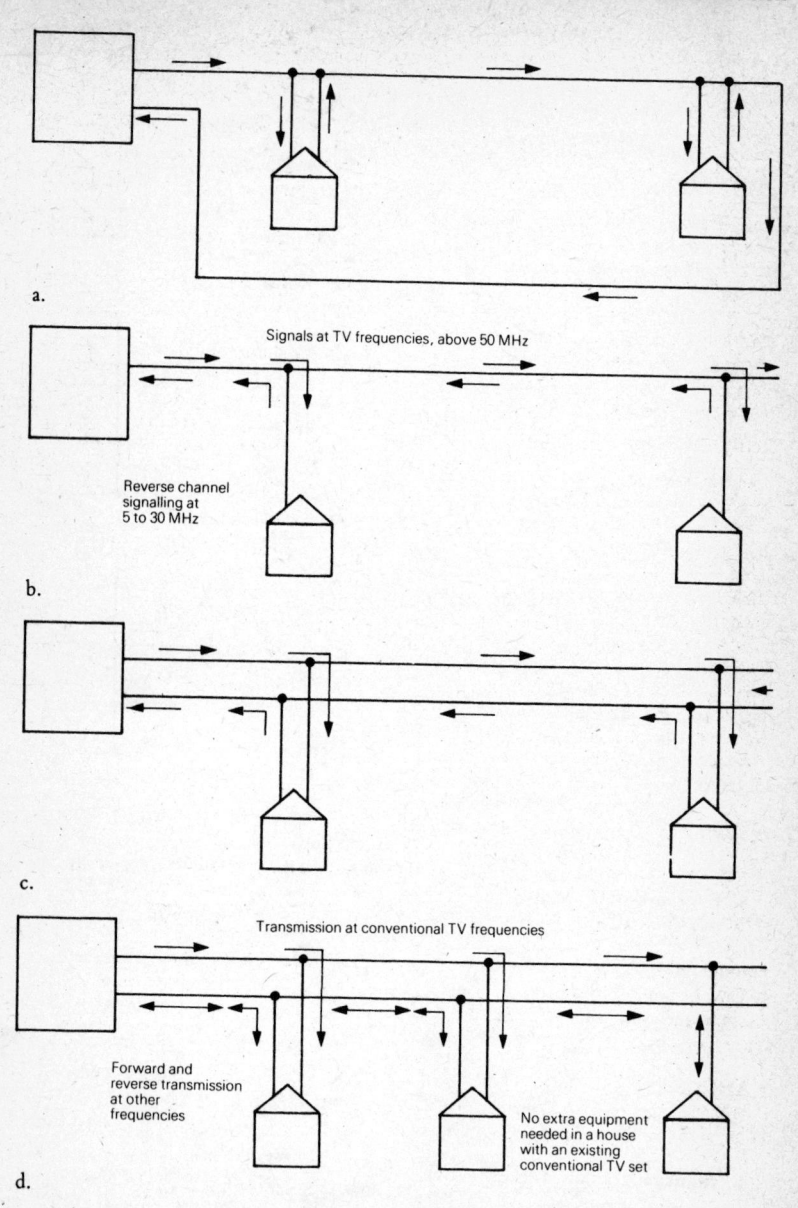

Signals at TV frequencies, above 50 MHz

Reverse channel signalling at 5 to 30 MHz

a.

b.

c.

Transmission at conventional TV frequencies

Forward and reverse transmission at other frequencies

No extra equipment needed in a house with an existing conventional TV set

d.

Reverse-channel signalling: illustrated are four different systems allowing two-way communication in CATV networks. a. A cable looping back to the cable head b. Two-directional transmission along a single cable c. Two separate cables d. Utilization of both a conventional and a two-way cable.

retrieval system (see *information retrieval systems*).

CAL computer aided (or assisted) learning.

call directing code a *code* directing messages between two communications *stations*.

calligraphic plotter a *plotter* which draws an image on a *CRT* consisting of lines only. It is used in *computer graphics*. (See also *raster plotter*.)

CAM 1. computer aided manufacturing systems. 2. computer addressed memory.

Cambridge Ring a *local area network* developed at Cambridge University, UK.

camera-ready copy *hard copy* manuscript (usually a typescript) of a quality and layout suitable for photographing and printing directly.

CAMIS Computer Assisted Make-up and Imaging System.

Cancerline a group of three *databases* produced by the US National Library of Medicine. They are CANCERLIT (a bibliographic database), CANCERPROJ (covering research projects in progress) and CLINPROT (Clinical protocols for treatment). Cancerline is accessible via a variety of *hosts*.

CANDOC a Canadian electronic document ordering service operated in conjunction with *CANOLE*.

canned paragraphs used in *word processing* to describe pre-recorded paragraphs which are in frequent use, and can be combined in a variety of ways.

CANOLE Canadian On-Line Enquiry. The *host* services of the National Research Council of Canada.

capacitance electronic disc see *video disc*.

capacitive video disc see *video disc*.

CAPS a printer's abbreviation for 'capital letters'.

CAPTAIN a Japanese *viewdata* (interactive *videotex*) system.

card (punched card) a card of standard size, thickness and shape used to input data and instructions. The most frequently used card is 7¾ inch by 3⅛ inch and has 80 vertical columns numbered from left to right. Each column has 12 possible punching positions, which accommodate the encoding of *characters*, one to a column. A numeric character requires only one hole to be punched in its column (these positions are numbered vertically 0-9), while other characters require two or more holes to be punched in the columns into which they are to be entered. The pattern of positions punched for any given character is determined by the *code* used by the *card punch*. Not all equipment uses the same codes, and consistency between the codes used by *card punch* and *card reader* therefore has to be checked before cards can be used to *input*. In the event of inconsistency, many computers offer facilities for automatic recording and repunching of cards (see illustration).

card column one of the columns (typically 80 altogether) on a *card* into which information can be punched.

card feed a device which moves *cards* one by one into a machine where they can be read.

card punch a device which perforates *cards* in specific locations under the guidance either of a computer, or of user at a *keyboard*.

card reader a device which permits the sensing of information punched on *cards* (by means of *brushes*), and then converts this information into electronic messages.

CARIS Computerised Agricultural Research Information System of the United Nations Food and Agriculture Organization.

carriage return a key, or character, which ends a line of type when activated and brings the cursor down to the start of the next line, eg in *phototypesetting*. The word has been adopted from typewriter terminology.

carrier signal a signal which carries packages, or streams, of information.

carrier wave an electromagnetic signal to which information can be added by means of *modulation*.

CAS 1. Current Awareness Service. 2. Computer Acquisition System.

CA search *database* containing *Keyword* and *Chemical Abstract* volume index entries and bibliographic citations. It is based on the entire contents of *Chemical Abstracts*, and therefore contains information which represents the sum of that within *CA Condensates* and *CASIA*. It is available via *BRS*, *Pergamon-Infoline*, *Lockheed* and *SDC*.

CAS files Chemical abstract service files: a

generic name for several files based on chemical abstracts: *CA Condensates, CA Search, CASIA, CIN*.

CASIA Chemical Abstracts Subject Index Alert: a *database* containing the volume index of *Chemical Abstracts*. When mounted *on-line*, *CASIA* records are usually linked to the main *CA Condensates* files.

cassette a portable container for film (videocassette) or *magnetic tape*.

cast-off a term used by printers to mean estimating the number of pages a manuscript will make when it is composed (see *composition*).

CAT 1. computer assisted teaching/train-

Examples of punched card before and after the code has been punched.

ing (see *computer assisted instruction* and *educational technology*). 2. computer-aided translation. Synonymous with *machine-aided translation*. 3. computer-aided typesetting.

catalog(ue) a list of items arranged for easy reference.

catastrophic error when so many or such large errors occur in a computer *program* that the *job* is terminated.

CATCALL Completely Automated Technique for Cataloguing and Acquisition of Literature for Libraries.

Cathode ray tube an electronic display device, similar to a television picture tube, used to display information including graphics. Its surface provides the screen in *visual display units* and *word processors*. The name is often abbreviated to CRT.

Cathode ray tube display unit a *visual display terminal* (VDT) in which a *cathode ray tube* (CRT) is used to display information and/or graphics.

CATNIP Computer-Assisted Technique for Numerical Index Preparation.

CATV community antenna television (see *cable television*).

CB *Citizens' band radio.*

CBPI Canadian Business Periodicals Index: a *database* covering Canadian business, industry, finance and related matters. Accessible via *QL* and *SDC*.

CCC *Copyright Clearance Center.*

CCD *Charge-Coupled Device.*

CCIR *Comité Consultatif International des Radiocommunications.*

CCITT *Comité Consultatif International Télégraphique et Téléphonique.*

CCLN Council for Computerised Library Networks, US.

CCTA *Central Computer and*

Telecommunications Agency.

CCTV *closed circuit television.*

CDC call directing code.

CDI Comprehensive Dissertation Index. A *database* covering US doctoral dissertations, accessible via *BRS*, *Lockheed* and *SDC*.

CED 1. Centro Elletronico di Documentazione. A Rome-based *host* linked with *Euronet Diane*. 2. capacitance electronic disc.

Ceefax the *teletext* system operated by the British Broadcasting Corporation (BBC) in the UK.

Celtic French telecommunications device to improve the efficiency of voice transmission by data *compression*. Uses *voice activation* to assign channels to users.

Central Computer and Telecommunications Agency agency set up by the UK Government to advise on public purchasing policies in the fields of *computers*, *telecommunications* and other *information technology*.

central office a US term for *telephone exchange*.

Central Office of Information (COI) UK Government agency with responsibility for information and publicity.

central processing unit the heart of a computer. It contains the *arithmetic and logic unit*, the core memory (see *memory*) and the control unit, which directs and coordinates the operation of the computer and its peripheral units. It thus carries out all the arithmetic, logic and control operations.

central processor see *central processing unit*.

CEPT the Conference of European Postal and Telegraph Administrations. Makes recommendations on telecommunications standards.

CETA Chinese-English Translation Assistance Group; operating a *pure MAT* system which aids translation from Chinese to English, but not vice-versa (see *pure*

MAT and *machine-aided translation*).

chain indexing the production of an alphabetic index in which each item appears under the title of each of the terms under which it falls within a *hierarchical classification*. Terms lower in the classification are listed after the higher terms. Thus, for example, an item on Shetland Ponies may have five entries:
Shetland Ponies p20-25
Ponies – Shetland Ponies p20-25
Horses – Ponies – Shetland Ponies p20-25
Mammals – Ponies – Shetland Ponies p20-25
Animals – Mammals – Ponies – Shetland Ponies p20-25.

chain search a method of searching which leads from one *record* to another until a required *record* is found, or the end of the chain is reached.

channel a pathway along which signals can be transmitted. Can be used in communications to mean a path for transmitting signals in one direction only (contrast with *circuit*).

character a single number (0-9), letter, punctuation mark, or other symbol, eg *, #.

character code a *code* used to represent a *character* in a computer, or in telecommunications.

character density the number of *characters* stored in a unit length, area or volume.

character-generator in a *CRT phototypesetter*, this is the device which forms the characters on output.

character reader a device which inputs printed *characters* into a computer (see *optical character recognition* and *magnetic ink character recognition*).

character recognition the use of *pattern recognition* techniques to identify *characters* (especially *alphanumeric*). There are several types of technique, eg *magnetic ink character recognition* (MICR), *optical character recognition* (OCR).

character set the collection of numbers, letters, graphics and symbols that can be generated by a particular system.

character spacing reference line a vertical line used to determine the horizontal spacing of characters (see *optical character recognition*).

character string a sequence of *characters*.

character stroke a line segment or mark used to form characters in *optical character recognition*.

character style the way in which a character is constructed in *optical character recognition*.

charge-coupled device (CCD) a *volatile storage device* made out of *silicon chips*.

check bit a *binary check digit*.

check character a *character* used in carrying out an error check, as with a *check digit*.

check digit when an item is catalog(ue)d, or classified, using a sequence of digits, a check digit is sometimes added to assist the automatic detection of errors in transcription. This check digit normally has an arithmetical relationship to the classification digits. A departure from this relationship therefore indicates that an error has been made in coding, or transcribing, the sequence of digits. (See, for example, *ISBN* and *ISSN*.)

Chemdex a *database* structured like a dictionary containing names and nomenclature derived from *Chemical Abstracts*. It is accessible via *SDC*.

chemical structure retrieval *information retrieval techniques* used to index and retrieve information about chemical compounds. There are three main methods of representing chemical structures for use with computers: a. fragment codes. Standard structural fragments of compounds are given codes, and these codes combined to describe the structure; b. topological methods. All the atoms in a compound and their bonding are represented by tables; c. linear notation.

The best known system of this type is *Wiswesser Line Notation*.

Chemline *database* taking the form of an *on-line* dictionary, produced by the US National Library of Medicine and accessible via a variety of *hosts*.

Chemname a *database* produced by the Chemical Abstracts Service which takes the form of an *on-line* dictionary of chemical substance names. It is accessible via *Pergamon-Infoline* and *SDC*.

chip a description of a single integrated circuit. It is usually in a package between 1 and 5cm in length, and having between 6 and 40 for external connections. The type normally found in computer systems is called a *logic* chip. *Analog(ue)* circuits, eg an audio-frequency amplifier, can also be made as chips.

CI 1. *chain index*. 2. cumulative index.

CICIREPATO Committee for International Cooperation in Information Retrieval among Examining Patent Offices (see *ICIREPAT*).

CICS Customer Information Control System: a software system used by American banks to give local branches access to central records of customer accounts.

CIM computer input microfilm. *Microfilm* used for high speed input of information into a *computer*.

CIN Chemical Industry Notes: a *database* containing business-orientated information covering the chemical processing industry. Accessible via *Lockheed* and *SDC*.

CIP Catalog(u)ing In Publication. The provision of bibliographic data on material in the process of publication. The data are included in Library of Congress and British Library *MARC* files.

CIRCA computerized information retrieval and current awareness.

circuit 1. the path round which an electrical current flows. 2. in telecommunications, a means of two-way communication involving 'go' and 'return' *channels*.

circuit switching a telecommunications term. Individual *circuits* are interconnected through successive *exchanges* to establish a continuous end-to-end connection which provides for transmission in each direction. (Contrast with *packet switching*.) Synonymous with line switching.

CIS Congressional Information Service. Built upon a *database* containing the working papers of the US Congress: it is publicly accessible via *Lockheed* and *SDC*.

CISI Compagnie Internationale de Service en Information. A Paris-based *host* linked with *Euronet Diane*.

citation a reference to a work from which a passage is quoted, or to a source regarded as an authority for a statement, or proposition.

citation indexing a *citation* here refers to the bibliographic reference made by one document to another. A citation index lists the 'cited' documents, usually arranged alphabetically by the author. Under each cited document, are also listed those documents which have cited it subsequently. Thus, from a known document, a searcher can find presumably related documents published more recently. The most developed and famous citation index is the *Science Citation Index*. (See also *co-citation indexing, bibliographic coupling*.)

Citizens' band radio (CB) is intended for the transmission of messages between individuals and groups, eg businesses. It uses low power transmitters, typically giving a range of up to 20 miles. In the US, where it is far more highly developed than elsewhere, most CB radios are installed in cars or trucks, and used in much the same way as radio telephones. CB radio has become so popular that there is now over-crowding of the *frequency band* allocated for CB radio (around 27 *MHz* in the US). American users of CB radio have developed a vocabulary of their own, which is currently being exported to other countries. CB radio can also be used for other purposes, eg the remote control of electronic devices.

CLAIMS/CHEM Class Codes Assigned Index Method Search/Chemistry. A US *database* containing US chemical and chemically related patents with equivalents for Belgium, France, UK, West Germany and the Netherlands. Accessible via *Lockheed*.

CLAIMS/GEN a *database* containing US general, electrical and mechanical patents. Accessible via *Lockheed*.

C-language a *microcomputer language* developed by Bell Laboratories in the US.

classification any systematic scheme for the arrangement of documents, usually according to their subject. (See, for example *Dewey Decimal Classification*.)

classification and coding systems conventions which provide a logical and meaningful basis upon which to code information or artefacts. The aim is to identify items in a way that facilitates easy identification and selective access and retrieval. (See, for example, the *Brisch*, *Dewey decimal*, *NATO*, *Opitz*, PERA and *Pittler* systems.)

clock an electric pulse generator which synchronizes all the signals in a computer.

close classification arrangement of subjects in a *classification* system involving a number of small subdivisions. This is necessary for the adequate definition of documents in a specialist collection (see *classification*).

closed circuit television (CCTV) a form of *cable television* accessible to a limited user group. It is currently used especially in security systems and in educational applications (see *Educational Technology*).

closed loop system, or form, of control in which there is automatic feedback.

closed user group this term appears mainly in the context of *viewdata/videotex*. It refers to a group of users who are allowed access to data or information which is not made available to other users of the system.

cluster 1. a group of *terminals* and other computer devices connected so that they operate together. 2. a mathematical process whereby objects are grouped together as members of clusters. The members of each cluster have properties in common which distinguish them from other clusters. Used in some computer *information retrieval* and classification systems.

cluster systems see *shared resource*.

CM communications multiplexor (see *multiplex*).

CMC code for magnetic characters (see *magnetic ink character recognition*).

C MOS complementary metal oxide semiconductor. A *transistor* used in *integrated circuits*. Employed where low power consumption is required. (See also *N MOS* and *P MOS*.)

CNC *computer numerical control*.

CNI Canadian News Index. A *database* giving coverage of current affairs in Canada. It is accessible via *QL* and *SDC*.

COAM equipment customer owned and maintained communication equipment – such as terminals.

coax an abbreviation of *coaxial cable*.

coaxial cable a communication cable consisting of an inner central conductor, usually of copper, insulated from an outer conductor, also usually of copper. When high frequencies are passed down such a channel, there is very low loss of energy. Several such cables can be combined into a single bundle.

COBOL Common Business Orientated Language. A *high level programming language* designed especially for the manipulation of business data. It uses terms which are related to ordinary English words.

co-citation indexing a development from *citation indexing*. Co-citation pairs documents which have been cited in common by other documents. The method rests on the assumption that the more frequently documents are found to be cited in common by other documents, the more likely it is that there is a subject relationship between the documents. (See also *bibliographic coupling*, *information retrieval techniques*.)

Codabar code a type of *bar-code*.

CODATA Committee on Data for Science and Technology of the International Council of Scientific Unions.

code a *machine language* representation of a *character*.

code conversion different terminals may use different *codes*. Code conversion is therefore necessary before these can communicate with each other. For example, a terminal using *ASCII code* can only be connected to one using *Baudot* code, if *digital* devices in the communication network perform the conversion. If different *line control procedures* are used, these will also need conversion before communication can take place.

code extension the extension of *character codes* to cover a greater range and variety of characters than are covered by the standard codes.

code key a key on the keyboard of a *data processing* or *word processing terminal* which is used in giving instructions to the computer.

codec this is an abbreviation of coder-decoder: a device which converts analog(ue) signals to *digital* and vice versa. It is typically used to convert analog(ue) signals to digital form for *digital transmission*, and then, after transmission, to reconvert again to the original analog(ue) form.

coden a five-character code which uniquely designates the title of a periodical or other serial.

coding sheet a paper form printed with a grid of rows and columns. Characters can be entered into each box on the form. The resulting format makes it easier to transcribe the information into machine-readable form, eg at a *card punch* or *terminal*.

COI *Central Office of Information.*

COINS Computer and Information Sciences.

cold type any *typesetting* technique that does not require the formation of characters from *hot metal*. Synonymous with 'strike-on'.

collotype a photo-mechanical process of printing from a raised gelatine film image on a glass support.

colo(u)r bars standards designed to ensure that colo(u)r television and computer terminal equipment work well together.

colo(u)r blind film photographic film that is not sensitive to all colo(u)rs: sometimes used for *microfilm*.

colo(u)r microform see *microform in colo(u)r*.

COM normally *computer output microfilm* (sometimes *microform* or *microfiche*). Instead of producing paper output, COM systems reduce the same information to *microfilm*, thus offering a number of advantages over paper output: a. speed – pages are produced at speeds in excess of 2,000,000 lines per hour (cf 20,000 lines per hour for *printers*); b. economy – microfilm cost is 20 per cent that of equivalent paper costs; c. distribution and storage – clearly cheaper and easier, being less bulky. The main disadvantage is continuing reader resistance to the medium.

Comité Consultatif International des Radiocommunications (CCIR) one of the three main organizations within the International Telecommunications Union (ITU). CCIR is particularly involved in examining and recommending standards for long range radio communications.

Comité Consultatif International Télé-

graphique et Téléphonique one of the three main organizations within the *International Telecommunications Union* (ITU). CCITT is particularly involved in examining and recommending standards relating to telephonic and telegraphic communication.

command language language used in *on-line searching* to facilitate a dialogue between a user and a *host* computer. It consists of a restricted range of instructions and terms. Command languages vary from host to host, eg *Dialog* for *Lockheed* and *ORBIT* for *SDC* (see *on-line searching*).

common carrier an organization which provides communications services to the general public, eg British Telecom in the UK, or AT & T in the US.

common command language a *command language* established for the searching of more than one *host*. Common command languages are particularly valuable when host computers are connected in a *network*, eg *Euronet*. The searcher can then more easily switch from host to host during a search. Such a common command language is offered, for example, as an option by several *DIANE* hosts within *Euronet* (see *on-line searching*).

common software computer *programs* and *routines* which are in a language common to several computers and users.

Common Technological Policy an EEC policy dealing with cooperation between member states in various fields of advanced technology, including computer and information technology.

communicating word processors word processors connected via a network to allow very rapid office-to-office and/or institution-to-institution communication of text.

Communication Satellite Corporation (COMSAT) based in the US, COMSAT provides technical and operational support services for the transglobal satellite communication systems using the Intelsat satellite services.

communications mix a combination of communication media and/or techniques.

communications satellite an artificial satellite, usually in a *geostationary orbit*, which amplifies and converts the frequency of signals received from Earth stations. The resulting signal is then retransmitted back to ground-based receivers (see *satellite communication*).

Communication Technology Satellite Canadian *communications satellite* launched in 1976. The first operational satellite dedicated to direct television transmission (see *direct transmission satellite*).

community antenna television see *cable television*.

COMPAC computer output microfilm package (see *COM*).

compaction algorithm an *algorithm* to achieve data compaction (ie reducing data into a more compact form requiring fewer *bits*). It is used, for example, in digital *facsimile transmission*.

companding compressing and expanding. Information in *bits* can often be recoded into a more compact form for transmission, using a *compaction algorithm*. A complementary algorithm can recover the original form; the entire process represents companding.

compatibility refers generally to the ability of two (*hardware/software*)devices to work in conjunction: eg if a *floppy disc* can be read by a particular *word processing* system, they are said to be compatible. Computer compatibility usually means software compatibility.
If a *program* can be successfully run on two computers, without alteration to the program, then the computers are said to be compatible. (See also *upwards compatibility*.)

COMPENDEX Computerised Engineering Index. A *database* covering all branches of engineering. Accessible via most of the major *hosts*.

compile to translate a *high level language* into a sequence of *machine language* instructions for the computer.

compiler programs which *compile*.

compose see *composition*.

composition 1. preparation of copy in a *format* which can be duplicated. 2. filing *records* in a *storage* device. (See also *computer-aided phototypesetting*.)

compression (techniques) see *compressor*.

compressor any electronic device which compresses the range of a signal. The aim is to improve the proportion of wanted to unwanted signal and enhance the efficiency of transmission (see *companding*).

COMPUNICATIONS computers and communications. A jargon term referring to the joint use of computers and communication systems. Thus it has a similar meaning to *information technology*.

computer an *electronic* device which receives *input data*, puts them into *storage*, operates on them according to a *program*, and *outputs* the result to the user.

Computer Acquisition System a US *tele-ordering* system.

computer-aided design the use of computers to aid design involves *computer graphics*, modelling, analysis, simulation and optimization of designs for production.

computer-aided instruction refers to the use of *computers* as 'teaching machines'. The computer presents instructional material, and asks questions of increasing difficulty, at a rate determined by the correctness of a student's responses. If a student is unable to give correct responses, the computer is *programmed* to give additional instructional material, and to ask less demanding questions. By this method, called *programmed learning*, tuition is adapted to the needs of the individual students. *CAI* systems often incorporate a facility for monitoring each student's progress, thus obviating some of the need for examinations. If, however, examinations are required, a CAI machine can be programmed to administer them (see *educational technology*).

computer-aided manufacture the use of computers and numerical control equipment to aid manufacturing processes. (See *numerical control*, *computer numerical control* and *direct numerical control*.) It can also include *robotics* and automated testing procedures.

computer-aided page make-up the use of computers to automate, or semi-automate, the formation of text and graphics into discrete pages. Automatic page make-up is easier for 'directory' type material, where acceptable page-breaks can be readily codified. Semi-automatic page make-up uses a *cathode ray tube* (CRT) to present text for manipulation by an operator.

computer-aided phototypesetting phototypesetting means, in general, the preparation of material for printing using an optical system. A phototypesetter has four basic components: a light source, a character (type) store, some sort of lens system, and a light-sensitive recording medium (see diagram).
Several *generations* of phototypesetter exist, each exhibiting an advance in technology (examples of each type are still in use). First-generation photocomposers were adaptations of traditional 'hot metal' casters. The metal melting pot and mould of the latter are replaced by a photographic unit. These machines were often called 'filmsetters'. Such devices are essentially mechanical in operation and slow. Second-generation phototypesetters are also electro-mechanical in nature, but faster. Characters are mounted on discs, or drums. Their speed makes it desirable that they should be run by remote control – for example, by computer. This makes them the earliest generation of computer-aided phototypesetters.
Third-generation phototypesetters are essentially electronic, using CRT character generators. These use characters stored in *digital* form, whose appearance can then be electronically manipulated. The final characters are displayed on a screen, which

is then exposed to a photosensitive medium. They can be very fast: for example, an advanced system may set up to 3,000 lines of newspaper column per minute.

Fourth-generation phototypesetters have a digital character store, but employ *lasers* for imaging, and have good potential for graphics as well as text.

The input devices of phototypesetting have changed considerably. Early types used '*blind keyboards*' for *punched tape* (ie operators could not see what was set until composition had taken place). *Direct-entry phototypesetters* are now common. A variety of outputs now exist: *paper tape*, *magnetic tape*, *floppy discs*, etc. Text can then be displayed (on hard-copy, or a screen) for examination and correction before being fed into the computer-controlled phototypesetter. *Word processors* can be used as the input stage to a phototypesetter: it is becoming increasingly easy to *interface* such devices, with consequent savings in costs of *keyboarding*.

The computer editing of text followed by input to a phototypesetter is known as 'text processing' (although this is sometimes also used as a synonym for word processing).

computer-aided translation synonym for *machine-aided translation*.

computer-aided typesetting a general term for the use of a computer at any stage in the typesetting, or composition, process. At the simplest level, it is concerned with automatic *justification*, hyphenation, etc. At a more advanced level, it refers to *word processor* input to the *phototypesetting* operation, or direct to a phototypesetter, as well as *computer-aided page make-up*.

Computer and Control Abstracts a UK-based *database* covering computers and control. This includes *cybernetics*, information science, mathematical techniques and *software*. Accessible via most of the major *hosts*.

computer-assisted instruction see *computer-aided instruction*.

computer-assisted interactive tutorial system a system in which a *computer* is programmed to perform the role of tutor in (normally) a one-to-one tutorial. The student uses the computer, either passively, as a recipient of *programmed learning*, or, actively, in a mode which permits the student to ask questions and the computer to respond (see *computer-assisted instruction*, *programmed learning* and *educational technology*).

computer-assisted (or aided) learning (CAL) 1. another name for a *computer-aided instruction (CAI)* system. 2. the receipt of 'tuition' from a *CAI* system.

computer-assisted teaching synonym for *computer-assisted instruction*.

computer bureau an agency which runs other people's work on its own computer and often offers other additional types of computing assistance and consultancy.

computer conferencing the interchange of messages on a particular topic via a computer *network* (see *electronic mail*, *electronic journal* and *teleconferencing*).

computer graphics the use of computers to generate and display pictorial images. A user can generate these images using either a *keyboard*, or some special graphic *input* device.

The simplest approach is that of *vector graphics*. If a keyboard is used, the parameters can be entered, and the corresponding curve then appears on the screen. More than one curve can be entered and displayed simultaneously, and a variety of colo(u)rs can be used. As well as two-dimensional shapes, three-dimensional objects can be shown in perspective. Once entered, these forms can be manipulated: they can be moved, elongated, rotated about any axis, etc. These operations can be effected in some systems by using a *light pen*, applied directly to the display screen. Other systems use a *graphics tablet*: a special stylus is used to draw and manipulate forms on a tablet of semi-conducting material. (See also *digitizing tablet* and *electro-acoustic tablet*.) An alternative to vector graphics is *raster graphics*. This utilizes a matrix of *pixels* covering the display screen; so that, when a particular group of these picture cells are illuminated, they describe an image. Each pixel has its own *code* and is switched on, or off, according to the controlling *program*. Using either vector or raster graphics, images, once entered, can be directed to a *storage device*, or transmitted to a distant *terminal*. In addition, the images can be directed to an *output* device to produce a wide variety of types of film, or print-on-paper copy (see, for example, *film recorders*, *flat bed plotters*, *graph plotters* and *ink jet printers*).

With this flexibility, computer graphics clearly has major implications for *computer-aided design*, animation techniques, the production of audiovisual aids to communication and the electronic communication of information in graphical form (as an alternative to text or numerical form). However, the current comparatively high costs of computer graphics systems may slightly delay their widespread introduction.

computer input microfilm see *CIM*.

computer journal see *electronic journal*.

computer language 1. *machine language*. 2. a language in which instructions are given to a computer by a programmer, or user.

computer letter a letter of standard form into which personal details (ie recipient's name and address) are inserted using *word processing* software. Used extensively in marketing.

computer mail see *electronic mail*.

computer memory see *memory*.

computer network see *network*.

computer numerical control describes a situation in which a number of *numerical control* machines are linked together via a *data transmission network* and thus brought under the control of a single numerical control machine.

computer output microfilm see *COM*.

computer printout see *printout*.

computer readout see *readout*.

Computer Software Copyright Act a US Act, passed in November 1980.

computer typesetting see *computer-aided typesetting* and *computer-aided phototypesetting*.

computing amplifiers see *operational amplifiers*.

computing power a term referring to the speed with which complex operations may be performed in a computing system.

COMSAT see *Communication Satellite Corporation*.

COMSTAR *communications satellites* provided by COMSAT General (a subsidiary of *COMSAT*) for *AT&T*, and used for internal communication within the US.

COMTEC Computer Micrographics and Technology. A Californian-based group of users and manufacturers of *COM* equipment.

concentrator a device which divides one or more data transmission channels into a larger number. Use of the latter is allocated in such a way as to maximize the throughput of data. Incoming data may, for example, be placed in a *buffer* to achieve this aim.

CONF see *Conference Papers Index*.

Conference Papers Index (CONF or CPI) *database* covering papers presented at about a thousand conferences per annum. It is accessible via *Lockheed* and *SDC*.

configuration the layout of the *hardware* in a particular computer system.

conflation algorithm a computer procedure (see *algorithm*) of particular value as an *information retrieval technique*. Where use is made of *natural language* terms for *indexing*, problems can arise from different forms of related words, eg 'computing' and 'computational'. A conflation algorithm reduces all words with the same root to a single form by removing all the derivational, or inflectional parts.

Confravision *teleconferencing* service offered by *British Telecom*. Two or three conference studios in different cities can be simultaneously interconnected.

CONSER Conversion of Serials Project. A US project to create and maintain large *machine-readable databases* on serials, eg journals, periodicals.

console that part of a *data-processing system* which allows the operator to communicate with the computer, usually utilizing a keyboard. (See also *terminal*.)

constant length field an entry on a document or card requiring a fixed number of *alphanumeric characters*.

Consultative Committee on International Radio (CCIR) one of the three main organizations within the *International Telecommunications Union* (ITU). CCIR is particularly involved in examining and recommending standards for long-range radio communication.

contact video disc see *video disc*.

content addressed memory see *associative storage*.

continuous tone used in graphics to describe a picture with continuously varying grey tones, as in an ordinary black-and-white photograph. (Contrast *half-tone*.)

contrast in *optical character recognition*, the difference between the reflectance of two adjacent areas (normally black text and white paper).

control character a *character* whose occurrence in a particular context can change a control operation. A typical example is a character that can initiate *carriage return*.

controlled indexing language see *information retrieval techniques*.

controlled vocabulary a fixed list of terms used to index records for storage and retrieval. The use of controlled vocabulary is normally required in *on-line searching*.

control track in *video recording*, a track on the tape records pulses which synchronize the *head* and the drive for recording and replaying.

convergence describes the coming together of technologies which in the past have been regarded as relatively distinct, eg computers, telecommunications, printing and publishing, to provide integrated systems. One example is the concept of the '*electronic office*'.

conversational language natural language used to communicate with a computer in *conversational mode*.

conversational mode *on-line* interaction between a computer and user in the form of a dialogue. Each 'participant' responds in turn to the information or response presented by the other.

Converse a system for the *on-line* description and retrieval of *data* using *natural language*. (Developed by *SDC* in the US.)

converter a computer *peripheral* which converts *data* from one physical form to another, eg *punched card* to *magnetic tape*.

copy a manuscript that is to be *composed* and printed.

Copyright Clearance Center a US non-profit-making organization which offers licensing arrangements for the *photocopying* of documents.

core memory see *memory*.

correcting code see *error-correcting code*.

COSAP Cooperative On-line Serials Acquisition Project (see *teleordering*).

COSTAR conversational on-line storage and retrieval (see *on-line*, *on-line searching* and *information retrieval systems*).

CPI *Conference Papers Index*.

CP/M Control Program/Microcomputers, a widely used microcomputer *operating system*.

CPM 1. cards per minute (the rate at which a *card reader* operates). 2. *critical path method*.

CPS 1. *characters* per second. 2. cycles per second (= hertz).

CPU *central processing unit*.

CPW *communicating word processor*.

crash the shutdown of a computer system due to a malfunction of hardware, or software. (See also *program crash*.)

Crecord *database* giving comprehensive coverage of the Congressional Record: the official journal of US Congress proceedings. The database is accessible via *SDC*.

CRESS Computer Reader Enquiry Service System (for libraries).

critical path method (CPM) a management technique for scheduling and controlling large projects; particularly those which involve a large number of interdependent phases. The nature of each phase and its dependencies is incorporated into a 'network' of events, and *software packages* are widely available for the mounting of such networks within a computer. This allows progress to be monitored, 'automatic' progress reports to be written, problems, eg hold-up and resource acquisition difficulties, to be analysed, and management strategies to be simulated. Sometimes also called *PERT* or *network planning*.

CROSSBOW Computerised Retrieval of Organic Structures Based on Wiswesser. A *software* system for use with the *Wiswesser Line Notation* system to allow computer searches for information on chemical compounds.

cross fire *interference* between *telegraph* and *telephone* circuits.

crossfoot to add across several domains of numerical information.

cross talk in telecommunications, the unwanted transfer of energy from one circuit to another.

CRT *cathode ray tube*.

CRT composition a *phototypesetter* in which the characters are generated on the face of a *CRT*, rather than projected from a master grid or disc.

CSIPR Comité Special et International sur les Parasites Radiotélégraphiques (see *International Special Committee on Radio Interference*).

CSIRONET Commonwealth Scientific and Industrial Research Organization (CSIRO) Network. A *computer network* offering *on-line* access within Australia to the *databases* of CSIRO. (See *on-line searching*. See also *MIDAS* and *AUSINET*.)

CTI Centre de Traitement de l'Information: a Belgium-based *host*.

CTS 1. computer typesetting (see *computer-aided typesetting*). 2. *communication technology satellite*.

cuetrack in *video recording*, a track on the tape on which are recorded verbal instructions and editing codes.

CUG *closed user group*.

CULT system Chinese University Language Translation (Hong Kong). An *HAMT* system used to translate Chinese journals into English (see *HAMT*, *machine translation* and *machine-aided translation*).

cumulative index an index containing all items appearing in a number of separate indexes.

current awareness service any service that alerts users to new information likely to be of interest to them. Such information is typically bibliographic and retrieved by computers (see *selective dissemination of information*).

cursor a light indicator on a *VDU*, which shows where the next *character* is to be generated. The cursor can be moved across the screen by use of a key on the *keyboard*.

Cybernetics theory of communications

and control which accounts for the operations of systems in terms of *feedback* effects.

cycle a complete sequence of operations, at the end of which the series can be repeated.

cycle time the time required to read a *word* from the computer *memory* and write it back again. Also used in the more general sense of *response time*.

Cyclops audio graphics system a system originally designed by the Open University (UK) to augment the teaching of its (home based) students. Sound and graphics are recorded on an *audio cassette*. This can either be viewed 'locally', or transmitted over telephone lines (and via a *modem* and audio-graphic *terminal*) to be received on a television set. The tape can be stopped at any convenient time to allow tutor and students to confer: messages can be sent by drawing with a light pen on the television screen. The audio-graphic 'studio' and 'terminal' are now generally available, and their use in *distance teaching* and industrial training is being evaluated (see *educational technology* and *computer-assisted interactive tutorials*).

cylinder scanning a form of *scanning* used in *facsimile transmission*. So called because the object image, eg a printed page, is wrapped round a rotating cylinder which is scanned by the photosensitive device. Also called drum scanning.

DAA *direct access arrangement.*

DACOM datascope computer output microfilmer. An early output device for *COM*.

DAC System *Data Acquisition and Control System.*

daisy wheel printer a printer where the typehead is circular, with the characters attached round it on the ends of stalks. Such printers are commonly found as a part of *word processing* systems.

DAR daily activity report. A computer-generated report on *library* operations.

DARE 1. data retrieval system for the social and human sciences run by *UNESCO*. 2. documentation automated retrieval equipment. Equipment used in *automated information retrieval* and *document delivery* systems.

dark-trace CRT a *CRT* in which the surface does not glow brightly under electron bombardment. Instead, it is coated with a substance which produces a dark

image against the white surface of the tube.

DASD direct access storage device (see *direct access storage media*).

D

Daspan the data communications *network* facilities of the US-based multinational corporation *RCA*.

data groups of *characters* (*alphanumeric* or otherwise) which represent a specified value or condition. Data provide the building blocks of information.

data acquisition and control system a system in which a central *computer* is connected to a number of *remote terminals*. The computer receives data from, and transmits data to, such terminals while operating in *real time*.

databank usually an alternative term for a *database*. However, it is sometimes used to refer exclusively to a collection of factual, or numerical, data, as distinct from a *bibliographic database*, which gives references to documents.

database a store of data on *files* which can

Typical daisy wheel printer.

be made accessible to a computer. It is designed for operation in connection with an *information retrieval system*. The word is often hyphenated (data-base), or spelt as two separate words (data base).

Data Base Index an internal *SDC* index, acting as a master index to all SDC *databases*. It is used to aid the selection of the most suitable database for a given search.

database management system *software* which permits the monitoring of a *database*. For example, a check can be kept on who uses the database (security), what the level of usage is, and thus the database can be automatically updated.

data capture a general term covering techniques for converting data into *machine-readable* form. These include direct input on a computer, or *word processing*, keyboard, or via *direct acquisition* devices.

data carrier a medium, eg cards, paper, *magnetic tape*, or *discs*, used for recording data.

Datacentralen Danish *on-line* information system connected with *Euronet-Diane*.

DATACOM 1. *Western Union* data communication service linking over 60 US cities. 2. a global communications *network* used by the US Air Force.

data compaction refers to methods used to reduce the space and time required for data storage and transmission.

data compression reducing the size of data elements by changing the way in which they are *coded*.

data concentrator see *concentrator*.

data display unit a *display* unit based on a *CRT*.

Data-Inform A/S a Danish *on-line* information service containing information on travel and tourism. The *database* can be searched in any one of six languages, and confirmation and invoices can be effected automatically. The service is available via

Euronet DIANE information services.

data network a telecommunications *network* linking *terminals*, via which data are communicated.

data packet data divided into packets of about a hundred *bits* for transmission (see *packet switching*).

Datapac Network a commercial *network* linking *TELENET* in the US with the Trans Canada Network (ie the major American and Canadian computer networks).

Dataphone Digital Service *digital* data transmission system operated by *AT&T*. It is available in over 150 cities in the US.

data plotter synonymous with *plotter*, also sometimes called an *X-Y plotter*.

data processing (DP) includes all clerical, arithmetical and logical operations on *data*. Data processing in the context of information technology always implies the use of a computer for these operations.

data processing cycle the sequence of operations commonly associated with data processing (ie collection of data; conversion into *machine-readable* form; checking, or *validation*; processing, or manipulating the data; display and storage of results).

data processing system the computer *hardware* and *software* required to carry out *data processing* activities.

Data Protection Authority see *Lindop Committee*.

Data Protection Convention convention drawn up by the Council of Europe, laying down principles (similar to those proposed by the *Lindop Committee*) for the regulation of computer data to protect the privacy of the individual.

data radio the transmission of *data* using radio waves. An example is *teletext*.

data reduction transforming large bodies of raw *data* into useful, ordered, or simplified *information*.

data security the control of access to data held within a computer system. Usually achieved by issuing a series of confidential passwords to authorized personnel. A software system then checks that the correct sequence of passwords has been entered into the computer before it will output information, or permit stored information to be changed.

data set another name for a *modem*.

data set adapter device for *interfacing* a *computer* and a *modem*. It breaks down *bytes* from a computer into *bits* for *serial transmission*. This process is reversed for received signals.

Data-Star Swiss *on-line search* service offered by *Radio Suisse*.

data tablet a device with which to *input graphics*. A pen-shaped stylus is moved over a flat electromagnetically sensitive board, and the pen's position over the board is monitored by a computer. In this way, it is possible to draw images directly into the computer.

data tagging a technique in the compilation of bibliographic *databases*. Data tags are *codes* that indicate and uniquely identify specified types of *data* in a *source document*. The tags are attached to the bibliographic references, and then used as an aid in searching the database.

data transfer rate the rate at which data are written, read, or transmitted (normally measured in characters per second).

Datavision a Swedish *viewdata* (interactive *videotex*) system.

Datel services *British Telecom* services enabling data to be transmitted over the public switched *telephone network* (see *switching*) or *leased circuits*. There are six different Datel services, with varying *data transmission* rates and performance characters.

Datex-P the West German *public packet-switched data transmission network*.

db *decibel*.

DBAM database access method. A general computing term covering, eg *direct access* and *random access*.

DBI *database index*.

DBMS *database management system*.

DBOS *disc-based operating system*.

DC 1. *direct current*. 2. display console (eg *VDU*). 3. *digital computer*. 4. *decimal classification*. 5. data conversion. 6. detail condition (specification of condition). 7. design change. 8. direct coupled (see *direct coupling*).

DCR 1. *data conversion receiver*. 2. *digital conversion receiver*. 3. design change recommendation.

DD 1. *digital data*. 2. *digital display*. 3. *data demand*.

DDC 1. *Dewey Decimal Classification*, sometimes also abbreviated to *DC*. 2. direct digital control. Under the control of a *digital* computer.

DDCE *digital data* conversion equipment: equipment for converting digital data into some other form.

DDD *direct distance dialling*.

DDP 1. *digital data processor*. 2. distributed data processing (see *distributed processing*).

DDS 1. *digital display scope*. 2. *digital dataphone service*.

DDT 1. *digital data transmitter*. 2. a *software package* designed to assist the *debugging* of *programs*.

DE 1. *display* element. 2. *digital* element. 3. decision element. 4. display equipment.

deadlock in a computer *operating system*, deadlock exists when two programs, or processes, are concurrently executed, but each has allocated for its own use a resource which cannot be shared. Processing then ceases until priorities are allocated.

debit card system a form of *bank on-line teller system*. Plastic 'debit cards', incorporating *magnetic strips*, are issued by banks to their customers. The strip records the customer's personal identification number. To use the card, the customer inserts it into a *terminal* and keys in the corresponding identification number. Having thus gained access to the system, the customer can, via the *keyboard*, deposit or withdraw funds (or implement certain other types of request). The transaction is recorded on the card, as well as in the bank's computer.

debit magnetic strip(e) reader a device for reading the magnetic strips which appear on debit cards (see *debit card systems*), credit cards and cards used in *electronic funds transfer systems*.

debug (debugging) isolate and correct errors in a computer *routine* or *program*.

decibel a unit used in measuring the relative power of a signal. Usually abbreviated to db.

decimal classification system one of the most widely used systems for catalog(u)ing documents in libraries according to their subject matter. The main classes and sub-classes are designated by a number composed of three digits. Further sub-divisions are represented by numbers after a decimal point. For example, the number range 300-399 is allocated to Social Sciences; 369 = Sociology, 369.1 = Anthropology; 369.11 = Primitive Sex.

deck a set of *cards* used for a particular computer *program*. They are usually sequentially numbered so that they can be kept in order.

decode the interpretation by a computer of the instructions which are input to it (see *code*).

dedicated a *program*, procedure, machine, *network channel*, or system set apart for special use.

dedicated port an access point to a communication *channel* which is used only for one specific type of traffic.

default a particular value of a variable which is used by a computer system unless it is specifically alerted via the keyboard to use another value. For example, a computer chess player may assume its lowest level of play unless the operator keys in a higher.

definition the degree to which detail is shown by an image. Particularly significant in *display devices* and readers, eg *optical character reader*, *microform reader*.

deflection plates used to create *electrostatic deflection* in a *CRT*.

delay line a device for introducing a time delay into the transmission of data, eg for bringing together data arriving at different times. There are various types of delay line, eg acoustic, magnetic.

delimiter one of a pair of *characters* which mark the bounds, for example, of a *string* of characters.

delivery time the time interval between the start of transmission at an initiating *terminal* and the completion of reception at a receiving terminal.

demodulation the reconstitution of an original signal from a modulated one. The opposite process to *modulation*.

demodulator a device for carrying out *demodulation*.

densitometer an optical-electronic device used for measuring the density (ie degree of darkness) of photographic images.

derived-term indexing see *indexing*.

descender a typographic expression indicating that part of a lower-case character which extends below the normal body height, eg the lower part of the letters 'p' and 'q'.

descriptor a term, or terms, attached to a document to permit its subsequent location and retrieval. Descriptors are particularly employed in computerized *information retrieval systems*. (See also *keyword*.)

desktop computer an alternative name for a *microcomputer*.

destructive read-out when *data* are read from a *storage device*, and the record of the data in the *storage device* is then lost.

Dewey Decimal System a particular form of *decimal classification system*.

DFT diagnostic function test: a computer *program* to test the reliability of a system.

DG XIII Directorate-General (Section XIII) of the Commission of the European Communities (CEC) which deals with the information market and innovation. The Department for Scientific and Technical Communication forms part of DG XIII. It a. makes available research carried out under CEC patronage and b. promotes scientific and technical communication in the European community.

DGT the French *PTT* (Directorate General of Telecommunication).

diacritic an accent placed above or below certain letters in some languages; eg 'é' and 'ê' in French; 'ü' in German. Modern *phototypesetters* can include these accents on output, if they are identified in the input text *string*.

diagnostic program a *program* used to detect equipment malfunctions.

Dialog the *search language* and *software* used for *on-line access* to *Lockheed's databases*.

Dialorder a *document delivery system* operated by *Lockheed* in the US. A document, identified by an *on-line search* of a Lockheed *database*, can be requested from a *document fulfilment agency*. The request is made *on-line* but fulfilment is met by largely non-automated methods. (*Hard copy* is sent via the mail.)

Dialtech the *on-line search* service associated with use of the *ESA/IRS host* facility from the United Kingdom.

dial-up systems where *terminals* have access to a computer, via a *modem* attached to the telephone network, by dialling a telephone number for the computer.

DIANE an abbreviation of Direct Information Access Network for Europe. It refers to the information services offered over the *Euronet* system. There are currently over 300 hosts and 300 *databases*.

Didot (point) a typographic system of measurement used in continental Europe. Based on a *point* of 0.0148 inch (with twelve Didot points = one cicero).

DIDS *Domestic Information Display System*.

diffusing screen a translucent screen that evenly diffuses light. It is used in *microform readers*.

DIGICOM digital communications system.

digiography suggested term for processes involving digital storage of images, text and graphics, eg in *phototypesetting*.

digital representation of information by combinations of discrete *binary* units, as contrasted with representation by a continuously changing function: which is referred to as 'analog(ue)'. For example, a piano creates sound waves in discrete units of pitch, whereas the human voice can change pitch in a continuous manner. In computing, the unit used is usually a *bit*.

digital/analog(ue) converter converts *digital* signals into *analog(ue)* signals.

digital camera a camera which records images (*text* and *graphics*) in *digital* form. Digital cameras are used as *input* devices for *graphic information storage and retrieval systems*, and to record images for immediate transmission.

digital computer a calculating machine, normally electronic, which expresses all the variables and qualities of a problem in terms of discrete units (see *computer*).

digital optical recording the recording of *digital* information using optical (ie *laser*) techniques. DOR refers particularly to the *optical digital disc* (see *video disc*).

digital speech interpolation a method of transmitting voice signals more efficiently. The transmission channel is only active during the periods when the speakers are actually talking.

digital transmission the transmission of signals that vary in discrete steps with the input signal, rather than continuously. The steps are usually based on *bits*.
Digital transmission is used not only for data communication between computers, but also increasingly for the representation of continuously varying signals. The reason is that digital signals can be amplified at intervals along a long-distance channel without adding *noise*. This may be compared with *analog(ue)* transmission, where amplification of the signals can increase noise and, consequently, the error rate. An example of an area of new information

technology which currently employs both digital and analog(ue) devices is *facsimile transmission*.

digitize to convert a continuous variable into *digital* form.

digitized fount a *fount* stored in a *photo-typesetter* system in *digital* form.

digitizing tablet a type of *graphics tablet* (for which it is sometimes used as a synonym). (See *electro-acoustic tablet*. See also *computer graphics*.)

DIMDI Deutsches Institut für Medizinische Dokumentation und Information.
Performs abstracting and indexing of German medical literature for input to the *MEDLARS* and *MEDLINE* systems, and acts as a *host* to medical *databases* in West Germany.

din a multipin audio connector based on West German standards.

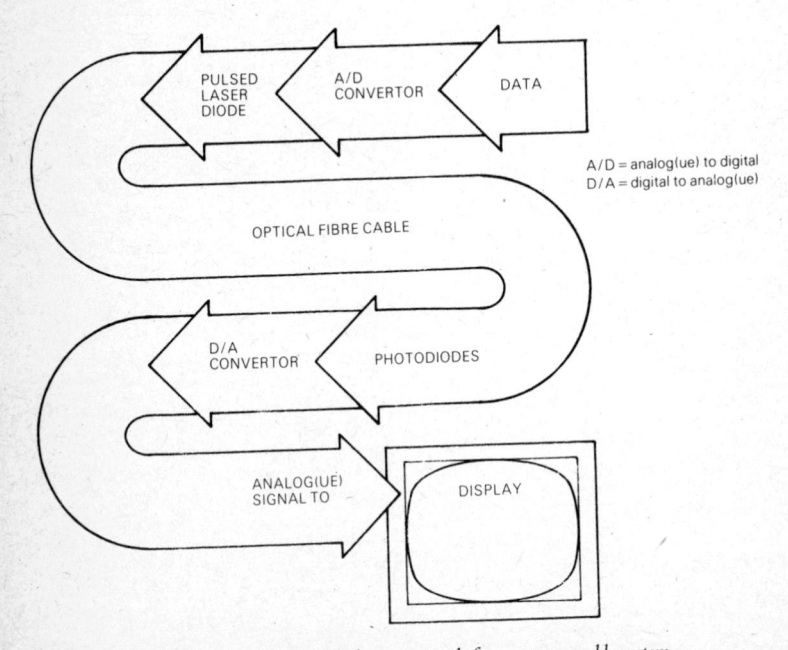

A/D = analog(ue) to digital
D/A = digital to analog(ue)

Digital transmission of pulsed light-wave signals for a one-way cable system.

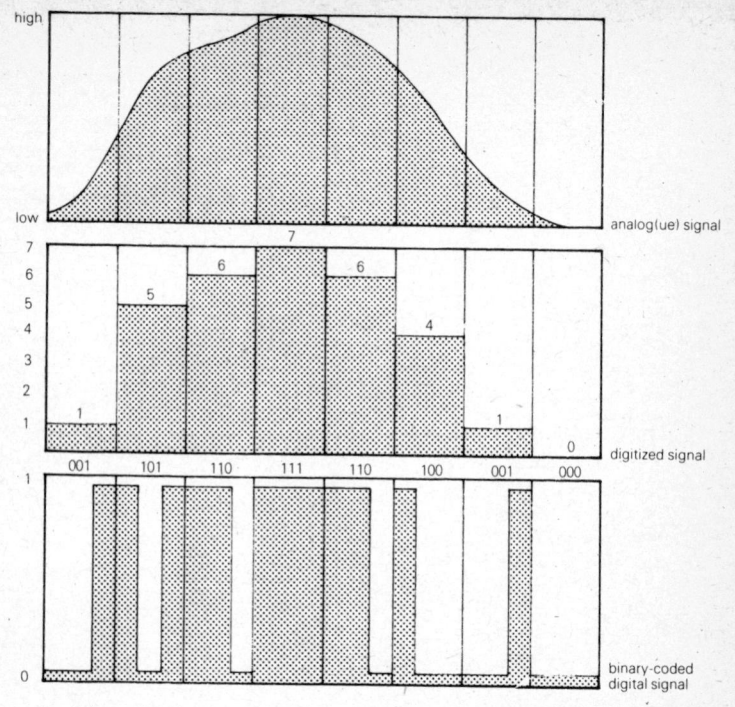

Breakdown of analog(ue) to digital conversion, the latter being transmitted in binary form.

DIN Deutsche Industrie Norm. A West German standards body.

diode an electronic device which permits current flow in one direction, but restricts it in the opposite direction.

DIODE Digital Input/Output Display System (see *digital*, *input/output* and *display*).

DIP Dual Inline Package: the most common form of *chip*.

direct access the ability to go directly to a desired item in a storage and retrieval system, without having to scan any other portion of the *storage file* first.

direct access arrangement (DAA) a device designed to protect the telephone network from high voltages, or large signals, which may be produced by the attachment to the network of consumers'

equipment. In the US, the *FCC* has ruled that equipment meeting certain standards can be attached to the network without a DAA.

direct access storage devices storage devices which provide *direct access* to the information required (see *direct access storage media*).

direct access storage media (DASM) media capable of storing *data* (and *programs*) in a form such that the time required to *access* particular elements is rapid and independent both of their location, and of the location of the last data element accessed (see *memory*).

direct coupling a way of connecting electronic circuits or components so that the amplitude of currents within each are independent of the frequency of those currents.

direct current (DC) a unidirectional current of effectively constant value.

direct distance dialling an automatic exchange service which enables a telephone subscriber to make calls to telephones beyond the local area.

direct entry photocomposition see *phototypesetting*.

direct-entry phototypesetter a typesetter in which a keyboard is incorporated into the photo-setting unit (see *phototypesetter*).

direct numerical control describes a situation in which a number of *numerical control* machines are linked together via a *data transmission network*. They can then be put under the direct control of a central computer with, or without, the guidance of a human operator. If there is no human operator, the system is often described as *computer numerical control* (CNC).

direct output *on-line output*.

direct plate exposure a system for the automatic processing of a printing plate, where the processor is attached to the exposure unit.

Direct Read After Write see *DRAW*.

direct transmission satellite a *communications satellite* which transmits messages to individual (ie home or office-based) receiving sets. Direct transmission is currently being investigated for use in television broadcasting.

direct voice input the input of information into a device, eg a computer, directly using the human voice, without an intermediate stage of *keyboarding* (see *speech recognition*).

DIRS DIMDI Information Retrieval Service. Information retrieval system operated by *DIMDI*.

disc see *magnetic disc*.

disc-based operating system an *operating system* in which *software* is held on one, or more, *magnetic discs*.

disc drive 1. a device which reads from, or writes to, *magnetic discs*. Also called a *disc unit*, or *magnetic disc unit*. 2. the mechanism within a *disc unit* which effects the necessary movements of the magnetic disc.

disc operating system a *program* which controls the operation of all activities related to the use of *magnetic discs* in a computer system.

Discovision an optical *video disc* system introduced in the US.

discretionary hyphen a hyphen inserted in a word by an operator to indicate the best place for a break if required by *justification*. If in subsequent setting it appears that the hyphen is not needed, it is removed from the output text.

disc unit a device which reads from, or writes to, magnetic discs. Also called a *disc drive* or *magnetic disc unit*.

dish (antenna) a transmitting or receiving *aerial* shaped like a dish. Typically used to receive radio and television signals from a *communications satellite*.

disk see *magnetic disc*.

diskette normally refers to a 5¼ inch *floppy disc*. Also called a *mini floppy* (see *floppy disc*).

display the production of a visual record on a television (or similar) screen. Can also be applied to the screen itself.

display background in *computer graphics*, the portion of a display image that cannot be changed by the user of the system.

display type large type (18 *point* or more) used for headlines, etc.

distance teaching instruction where the teacher and student are not in face-to-face contact. They communicate with each other by correspondence, *radio*, *television*, *CCTV*, *computer-assisted interactive tutorials*, etc (see *educational technology*).

distortion an undesired change in the form of a signal that may occur between two points in a transmission system.

distributed logic systems where *logic*, or *intelligence*, is distributed in the system

rather than located centrally. For example, some *word processing* systems link *intelligent terminals*, which may make shared use of other resources, eg storage, printer. (See also *shared logic*, *shared resources*.)

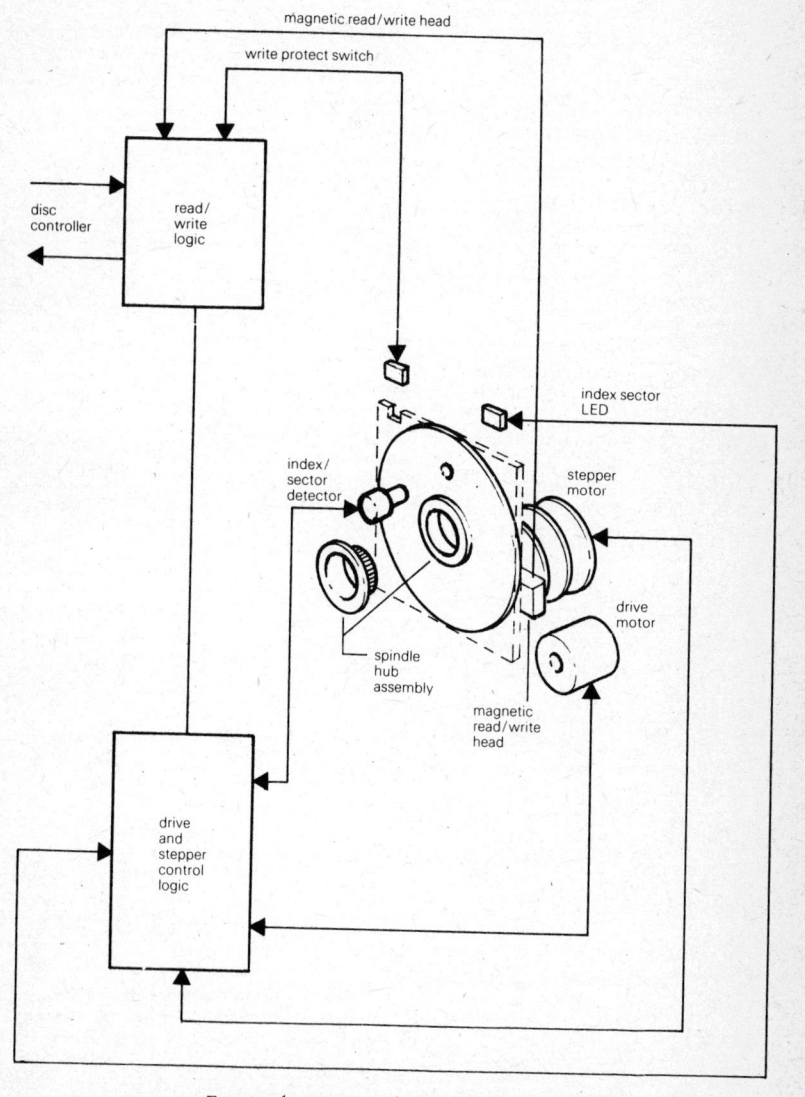

Functional components of a (mini) disc drive system.

distributed processing processing of data at different physical locations in a *distributed system*.

distributed switching the use of small *switching* units near to subscribers' homes in some types of *cable television* system. Such units need to be smaller, but more numerous, than conventional telephone exchanges.

distributed system a computer system in which several interconnected computers share the computing tasks assigned to the system.

DJNR *Dow Jones News/Retrieval*.

DMA direct memory access: a method for transferring data between computer *memory* and *peripheral units* without going via the *CPU*. This increases speed and efficiency.

DMM Defense Market Measures (system). A *database* recording all US Department of Defense contract awards. It can be searched via *Lockheed*.

DNC *direct numerical control*.

Docfax jargon for *facsimile transmission*.

document a medium and the data recorded on it. Most commonly refers to print on paper.

document assembly in *word processing*, the integration of different documents (or parts of them) into a single document.

documentation a permanent record of the way in which a computer system is to be operated. It includes, for example, *program* specifications and operating instructions.

documentation book a collection of all the documentation relevant to a particular *program*, or system.

document delivery (system) most current document delivery systems rely heavily on manual methods of storage, retrieval and distribution. Documents are

held as print-on-paper or *microfiche*, and distributed *on-demand* (normally in response to an inter-library loan request). Although electronic systems are currently being developed to satisfy each of the functions required of a delivery service, no large-scale, fully electronic systems are yet in operation. Some small experimental systems have been tried. For example, a range of articles indexed and abstracted for the *INSPEC database* (which can be searched *on-line*) are held as full (ie complete) text on *magnetic tapes*. These articles are on the database, and can be retrieved by the searcher. By switching to the full-text *file*, the relevant document is made immediately available on the user's terminal. Such integrated systems will become more widely available in the future, as a back-up to the on-line bibliographic search facilities currently available (see *on-line searching*). The European Commission is currently seeking the development of an integrated system (codenamed *ARTEMIS*) to operate in conjunction with *Euronet*.

At present, the major application of new technology to document delivery is in the 'request' phase. This stems partly from the introduction of fully automated document request systems in the inter-library computer *networks*. It also results from the activities of the major US *on-line database hosts* (*Lockheed* and *SDC*). They have introduced *switching* mechanisms which permit *on-line* requests to be made for documents identified during the course of a search (see *Dialorder* and *Electronic Maildrop*).

Electronic systems for the storage and retrieval of full-text documents are developing rapidly. For example, in the *Adonis* system, full-text articles are to be stored on *digital video discs*. Currently, this system will print out text for distribution by conventional means. In the future, digitized text and graphics will be transmitted to distant points for local print-out and display. The viability of this approach will be enhanced by developments in telecommunications – in particular, the more extensive use of *satellite transmission* and high capacity *broadband optical fibres*. The range of documents available via delivery systems will be increased through the use of ever larger capacity storage devices (probably *digital optical discs* and, perhaps, *holograms*).

document fulfilment agency any body which provides copies of documents requested by users. The requests may be generated as a result of *on-line searching* and transmitted to the agency via a computer *network*.

document retrieval systems for indexing, searching and identifying documents, from which information is sought (see *information retrieval systems*).

docuterm a name designating a segment of *data* in a way which indicates the content of that segment. The 'docuterm' is used to assist in subsequent data retrieval (see *information retrieval systems*).

Dolby noise reduction system used on audio tapes.

Domestic Information Display System a US *computer graphics* system which displays demographic data for the whole of the US in the form of colour-coded maps. Developed by *AOIPS*.

DOMSAT an Australian *communications*

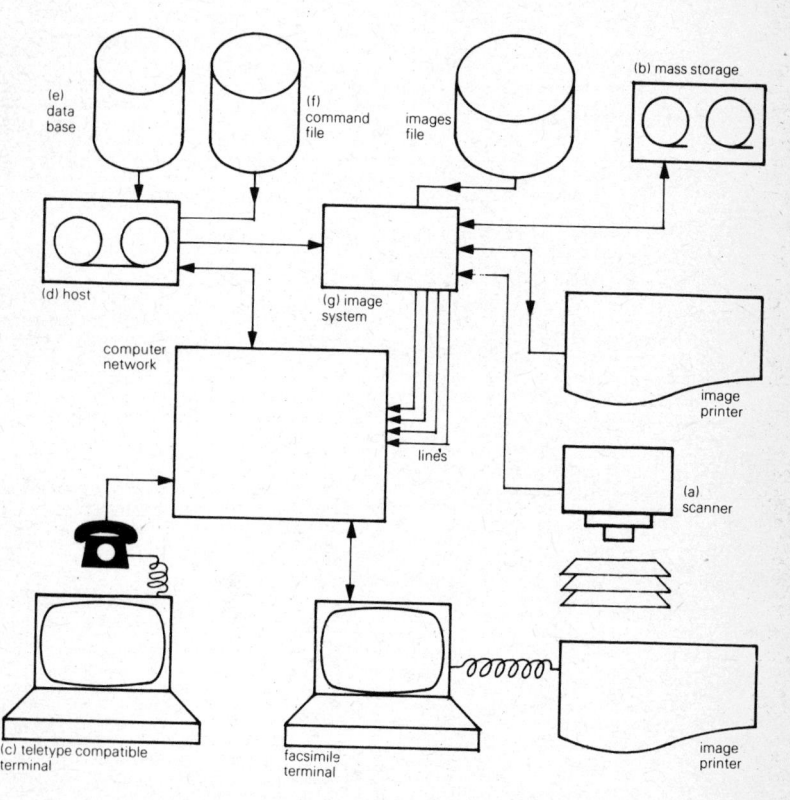

Processes involved in a typical document delivery system. Documents are digitized by an optical scanner (a) and transferred to mass storage (b) via the image system (g).By means of on-line searching, documents are ordered: (c) to (f). This order is transmitted to the image system, which forms a channel between the mass storage and the computer network. The user's document delivery terminal, connected to the network, receives and prints out the document via a facsimile terminal.

satellite system for domestic television and telecommunications.

DOR *digital optical recording*.

DOS *disc operating system*.

dot matrix a pattern, or array, of dots used for the presentation of *characters*, used in *visual display units* and in some printers.

dot matrix printer see *matrix printer*.

doubleword computer jargon for a *bit-width* of 64 bits.

Dow Jones News/Retrieval (DJNR) an *on-line information retrieval* service built upon a *database* covering business news, articles, etc, from the Dow Jones News Service, the Wall Street Journal, and other leading financial periodicals. The *database* is updated daily.

down has a specific meaning within the context of computer systems. A system is described as 'down' when it is not operating, due, for example, to a fault, repairs or maintenance.

downtime the period when equipment is not operating because of malfunction, maintenance, etc.

downward compatibility the ability of a more advanced system to interact with a less advanced one (often through an intermediary system).

DP *data processing*.

DPE *direct plate exposure*.

DPM 1. data processing manager (ie a person involved in managing data processing activities). 2. documents per minute.

DPS 1. *data processing system*. 2. Document Processing System: developed by IBM, US.

DRAW Direct Read After Write. With optical digital discs, information once written cannot be erased. However, the DRAW technique allows immediate identification of errors. These can then be corrected by rewriting data in a new section of the disc, and erasing the *address* of the incorrect section from the computer memory. The incorrect section will then always be ignored (see *video disc*).

Drexon DRAW Disc a *video disc* developed by Drexler Technology Corporation. Follows the Philips *DRAW* process, but the resultant disc is claimed to be less susceptible to deterioration.

drive any device used to load a disc (*magnetic disc*, *floppy disc*) on to a computer.

driver program *software* for a computer *typesetter* which provides the commands (interspersed with text *characters*) required for the operation of the typesetter in the correct format and *code* structure.

DRL Data Retrieval Language (see *information retrieval system* and *information retrieval techniques*).

DRO *destructive read-out*.

drop a connection between a *terminal* and a transmission *line*. For example, a 'subscriber's drop' is the line from a telephone *cable* to the subscriber's home or office.

drop line a *cable* which branches off from a feeder *line* to bring *cable television* into a subscriber's home.

drop out refers to faulty *magnetic tape* which prevents signals from being recorded.

DRS document retrieval system (see *information retrieval system*).

drum often used as an abbreviation for *magnetic drum*.

drum printer a type of *line printer*, which prints from a drum engraved with identical characters in each print position across the drum with the full character set engraved in each print position around the drum. (Synonymous with *barrel printer*.)

drum scanning a form of scanning used in *facsimile transmission*. So called because the object to be imaged, eg a page of print, is wrapped round a drum, which then rotates past the optical sensing device. Used as a synonym for *cylinder scanning*.

DSA *data set adapter*.

DSI *digital speech interpolation*.

dual-density disc a *magnetic disc* with twice the storage capacity of a standard disc with the same dimensions.

dual dictionary a printed *inverted file* of two identical parts. It is used in *bibliographic information retrieval* for the manual comparison of document numbers contained under each *descriptor*.

dual processor a computer system based on two CPUs. One is normally dedicated to information processing while the other deals with system operations.

dumb terminal a *terminal* with no independent processing capability of its own. It can only carry out operations when connected to a computer. (It is to be contrasted with an *intelligent* or *smart terminal*.)

dump the transfer of data from one computer *storage* area to another, or, more

usually, to *output*. Can also refer to the data obtained by this process.

duplex a method of communication between two *terminals* which allows both to transmit simultaneously and independently. A synonym for *full duplex* (contrast with *half duplex*).

duplexing the use of duplicate components, so that, if one fails, the system can continue to operate via the other.

Dvorak keyboard a form of typewriter keyboard designed to make the most frequently used keys most easily available. This can lead to improvements in typing speed and accuracy as compared with the standard *Qwerty* keyboard.

dynamic leading continuous motion (as contrasted with line-by-line advance) of the surface receiving images in a *phototypesetter*.

dynamic RAM *RAM* in which only one *bit* can be stored at an *address*, where it is held for a fraction of a second. (See, in contrast, *static RAM*.)

dynamic range the range from the weakest to the strongest signals that a receiver is capable of accepting as input.

EAM electrical accounting machine. An electronic machine which provides lists and totals from input data.

EAPROM electrically alterable programmable read only memory. Often used as a synonym for *EPROM*.

EAROM electrically alterable *read only memory*. Often used as a synonym for *EEROM*.

earth segment a term in *satellite communications*, referring to the Earth, or ground, *port*, as opposed to the satellite itself. Thus Earth segment costs are those of the ground receiving and transmitting stations, etc.

earth station a terminal which is able to transmit, receive and process data communicated by satellite.

EBCDIC (Code) Extended Binary Coded Decimal Interchange Code. A standard 8-*bit transmission code* for the exchange of *data* between items of equipment.

EBR *electronic beam recording*.

EBU European Broadcasting Union.

echo part of a transmitted signal reflected back with sufficient amplitude and delay time to be recognizable as interference. Echoes can be used to check the accuracy of data transmission.

ECHO European Commission Host Organization. This runs a referral and enquiry service, as well as acting as *host* to a variety of *databases*.

echo check method of checking the accuracy of data transmission by returning the received data to the sending point for comparison with the original message.

ECMA European Computer Manufacturing Association.

E-COM the US Postal Service's *electronic mail* system.

Econet a UK communications *network* which can be operated over a limited distance, and allows computers to share resources, eg printers. The system is primarily intended for schools and colleges, but can also be used in business.

ECS 1. *European Communications Satellite*. 2. *Experimental Communications Satellite*.

E-cycle see *machine cycle*.

edit the process of removing, or inserting, information by the intervention of an operator when a *record* is passed through the *computer*. Also used as an abbreviation for *text editing*.

editor besides the usual connotation, this term is used in information technology to mean: a. *software* which aids the *editing* of a *file*, normally by a user at a *terminal*; b. a *routine* which *edits* in the course of a *program*.

editorial processing centre/center (EPC) the concept of editorial processing centres first appeared in the 1970s. It refers to a system in which new information technology is shared by a number of journals. Such resource sharing is intended to produce savings in costs and to reduce the time-delays in publication experienced by conventional journals.
The key elements of the concept are to capture authors' manuscripts in *machine readable form* at an early stage; to avoid unnecessary re-typing; to use the computer for assistance in editing; to use *terminals* and *telecommunications* when communicating about submitted manuscripts; to introduce *computer-aided typesetting*; to monitor schedules more efficiently, and to aid financial management.
Studies of EPCs suggest that four types might be distinguished (minimum, intermediate, advanced and maximum) depending on the extent to which new technology is incorporated. The diagram shows a suggested 'intermediate' configuration, in which authors' manuscripts, and editors' and referees' comments are typed in an appropriate form to be read into a computer via an *optical character recognition* (OCR) device. In contrast, a minimum configuration makes less use of the computer, and the advanced and maximum configurations

E

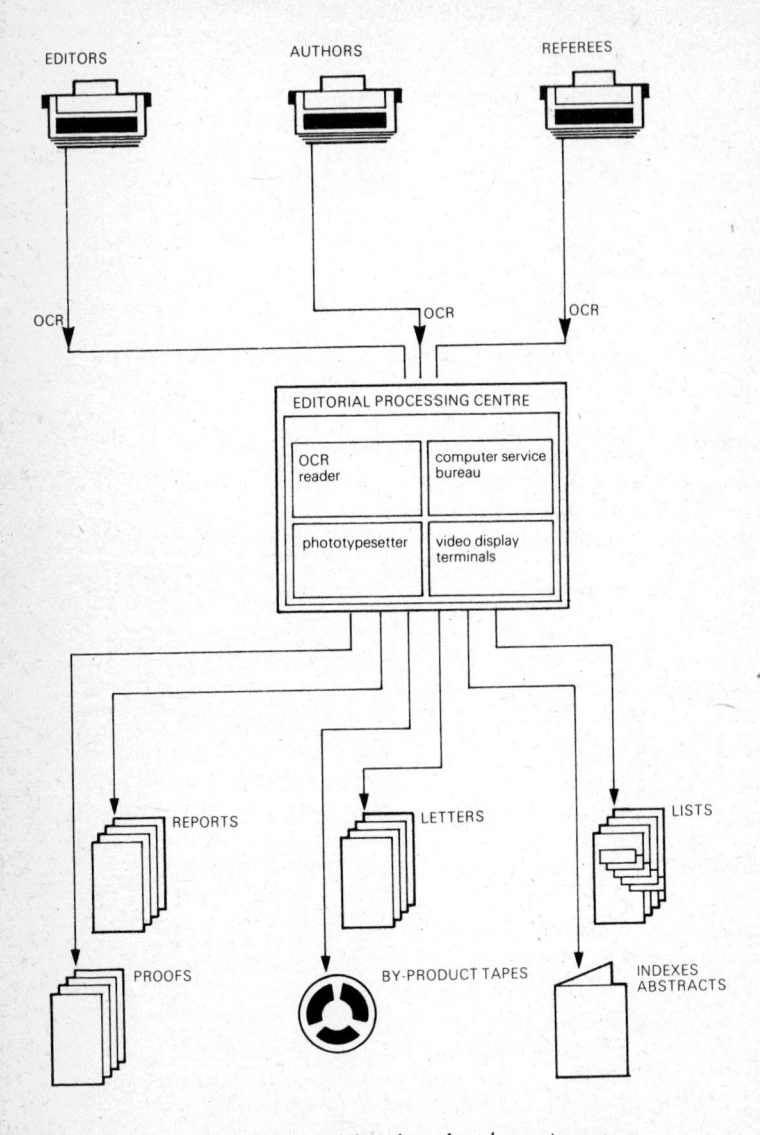

Flow of material through an editorial processing centre which employs a computer service bureau.

make use of a computer within the EPC with *visual display units* (VDUs).

EPCs have been the subject of a number of experiments, but do not actually exist in the fully integrated form described above. The phrase 'editorial processing centre' is now often used more loosely to describe any centralized editorial facility using new technology and responsible for the production of several journals.
(See also diagram opposite and *electronic journal*.)

EDP 1. educational data processing.
2. electronic *data processing*.

EDPE electronic *data processing* equipment.

EDPM electronic *data processing* machine.

EDPS electronic *data processing* system.

educational technology the development, application and evaluation of systems and techniques for improving the process of human learning. The term may refer to the social and psychological aspects of the learning process. But it is also applied to the development of devices and systems to assist: a. face-to-face teaching, eg films, tape-slide presentations, and *computer graphics*; b. *distance teaching*, eg broadcast lectures, *CCTV*, *video cassette* teaching packages and *computer-assisted interactive tutorials*; c. self-instruction, eg 'teaching machines' and *computer-aided instruction*. Educational technology and information technology have interacted extensively. Information technology provides the basis for much educational technology: but, equally, educational technology has contributed to such areas as the design of *user-friendly* systems. In particular, it has emphasized the use and design of *menus* as a means for gaining access to any required facility within a computer-based information system (see *computer-aided instruction* and *CYCLOPS audio-graphics system*).

EEROM electrically erasable read only memory. *Read only memory* which can be erased by passing an electrical current through it and then reused, ie new data entered. Often used as a synonym for *EAROM*.

effective time the actual time for which a piece of equipment operates.

EFT *electronic funds transfer*.

EHF extremely high frequency: radio waves above 30 GHz (see *spectrum*).

EHOG European Host Operators Group. A group representing the interests of European *hosts*, through which they can discuss their common problems.

EIES Electronic Information Exchange System (pronounced 'eyes'). An experimental computer *network* supported by the US National Science Foundation to evaluate the impact of information technology on personal communication. It connects members of professional groups to one another, offering them *electronic mail*, *teleconferencing*, and a facility to maintain and update personal electronic 'notebooks'.

elastic buffer (or **store**) a *buffer store* which can hold a variable amount of data. Such a store is often used in *digital transmission switching systems*.

ELECOMPS a *databank* of electrical components searchable *on-line* via *ESA/IRS*.

Electrical and Electronics Abstracts a UK *database* containing *abstracts* of papers published in all fields of electrical engineering and electronics.
It is accessible through most of the major *hosts*.

electro-acoustic tablet a type of *graphics tablet* in which the position of the pen, or stylus, is determined by measuring the time it takes pairs of pulses to reach the stylus' contact with the tablet surface. (See also *computer graphics*.)

electromagnetic delay line a *delay line* whose action is based on the time of propagation of *electromagnetic waves*.

electromagnetic wave electrical and magnetic vibrations at right angles to each other which travel together through space (see *spectrum*).

electron beam recording a *COM* output method, where a beam of electrons is directed onto a sensitive film.

electronic composition any computer-assisted method of *composition*, normally leading to output in page form.

electronic data processing *data processing* performed by electronic machines.

electronic document delivery systems see *document delivery*.

electronic funds transfer (EFT) a method for transferring funds from one account to another using computers and telecommunications. At least four types of EFT are currently in common use: a. transfers between computers at different banks; b. transfers between banks and other organizations, eg industrial firms; c. public access to *terminals* providing banking services, eg *machine readable* cards for obtaining money from cash dispensers; d. cards for making direct debit payments for goods and services via an electronic link.

electronic glass a transparent solid which is electrically conducting. It is used in a variety of information technology products, eg touch control *input* devices, using a finger instead of a *light pen*; visual display devices.

Electronic Information Exchange System see *EIES*.

electronic journal the 'electronic journal', in its simplest form, is the all-electronic counterpart of a conventional print-on-paper scholarly journal. The concept originated in experiments based on *computer conferencing* systems. The approach currently envisaged is based on a large *mainframe computer*, which acts as a central store.

The sequence of events might be as follows. An author prepares a research article at a computer or *word processing* terminal, and may then inform selected colleagues of its existence (using an *electronic mail* or message facility) inviting their comments. These colleagues 'call up' the article on their own *terminals*, and transmit comments to the author, either directly, or via a central electronic store. The author views these comments, and, if necessary, revises the article. The article is then submitted to an editorial office via a telecommunications link. The editor of the journal views the submission and, if appropriate, chooses referees to judge its acceptability for the journal. The referees carry out this examination at their own terminals. As with most conventional scholarly publications, the referees' comments may then be transmitted via the editor to the author – over the electronic network – and the article may be accepted, rejected, or revised. Up to this point, no user of the network, other than those chosen by the author and editor, has had access to, or knowledge of, the article. Once the article is accepted, it can be 'published' (ie made available via telecommunication links) to the subscribers to the electronic journal. They are alerted to its existence, and can call all, or part, of it up on their terminals. In this form, paper is not needed at any stage.

The flexibility of electronic communication means that the procedures of print-on-paper journals can be varied to determine which are best suited for the new medium. For example, each article might be structured so that the reader could then follow a variety of routes through its contents. Where a reference is made to other work, the user might be able to retrieve the work referred to immediately via a *document delivery system*. Again, the organization of the journal might be modified, with both editorial and refereeing activities more widely distributed among contributors. The electronic journal is increasingly being regarded as but one facet of the facilities offered by an electronic communication *network*. For a particular scholarly community, these may include: a newsletter, probably compiled centrally; a 'paper fair', where authors can input articles for comment from other participants; a message capability (person-to-person, or for all participants).

At present, experimental electronic journals exist, but none is either fully operational, or financially viable. If, and when, such journals do come into operation, it is likely that they will not be purely electronic, but will make provision for *hard-copy* at both input and output stages. Electronic journals may evolve from, and coexist with, the conventional journal publications system, possibly as a further development of *editorial processing centres.*

electronic mail a general term covering the electronic transmission, or distribution, of messages. Electronic mail can be distinguished from most areas of telecommunications by its capability for 'non-real time' use. Unlike a telephone conversation, messages can be transmitted at one time, for reception or reading at a later time. The delay can be brought about by the transmission system employed: for example, a 'store and forward' system which may have a central facility which stores received messages and subsequently transmits them down another line. Alternatively, messages may be stored at the receiving end, to be read at the convenience of the recipient. Most telephone networks are heavily used only at peak periods, and may be relatively little used at other times. A typical network may be only twenty per cent utilized. Electronic mail permits the usage of this spare capacity, transmitting in the 'off-peak' periods. This may be of particular value for business mail, much of which is already generated by computers for eventual insertion into other computers, eg orders, invoices, receipts, etc.

For the most part, electronic mail will involve the *digital transmission* of messages. However, the term can also include *facsimile transmission*, most of which still takes place by *analog(ue) transmission.* A combination of facsimile transmission and telex or *teletex* makes possible electronic mail systems able to handle both text and graphics. Electronic mail can be carried by a variety of devices – for example, by communicating *word processors* and transmission is not limited to telephone networks, since any telecommunications link can theoretically be used. Thus some see particular use for electronic mail over cable TV (*CATV*) links.

An electronic mail system which is controlled by a central computer, or *minicomputer*, and is intended for a limited set of users can be considered to be a *computer conferencing* system. *Electronic funds transfer* can also be subsumed under the electronic mail heading, since it may deal with the transmission of messages in 'non-real' time.

Electronic Maildrop a *document delivery system* operated by *SDC* in the US. A document identified by an *on-line search* of an SDC *database* can be requested, on-line, by means of a *switching* mechanism. The requests are currently met by fulfilment agencies who use largely non-automated methods (*hard-copy* is sent via the mail).

electronic message systems a general term, first used to describe communication via *terminals* in a communications *network.* It now covers a number of specialized services, such as *electronic mail, teleconferencing, videotex*, and communication between *word processors.*

electronic office a general term (equivalent to 'Office of the Future') used to describe a technologically feasible office environment which makes maximum, or optimum, use of information technology. Thus an electronic office would use electronic means for text and data handling, communication, information storage and retrieval facilities. (See also *work station.*)

electronic printer although the term could apply to any computer *printer*, it is used more specifically for devices which hold a *magnetic tape* record of the text and reproduce it via *digitized founts.*

electronics the design and construction of electrical circuits containing devices (such as *transistors*) whose operation depends on the behaviour of electrons.

electronic speech recognition see *speech recognition.*

electronic stylus an input device which allows images to be drawn. It can take the form of a *light pen* or a purely electronic device used in conjunction with, for example, a graphics tablet.

electronic switching system a *digital* telephone *switching* system which provides special services, such as speed dialling, call transfer and three-way calling.

electronic tutor a *computer* which provides *programmed instruction*.

electrophotographic printing there are two main types: a. indirect, where the photosensitive material is part of the *printer* mechanism and the image is transferred to paper; b. direct, where the photosensitive material forms part of the coated paper used as the recording medium.
In *laser* electrophotographic printing, *digital* information is fed to a laser which creates an image (in the form of a *dot matrix*) on a photoreceptor belt, or drum. This image is *toned*, transferred to paper and fused.

electrosensitive printing a method of printing where an electric current passes through a writing stylus to remove, or chemically modify, the top coating (often a wax) on a specially prepared paper.

electrostatic deflection deviating a beam in a *CRT* by means of an electrostatic field created by *deflection plates*.

electrostatic printing/reproduction a non-impact printing method whereby electrostatic charges are produced on paper in the design to be copied. Liquid or dry *toners* are attracted to the charged areas, which then become visible. Heat is used to fuse the toner to the paper (see *xerography*).

ELF extremely low frequency: less than 100 *Hz* (see *spectrum*).

Elhill *software* written originally for the US National Library of Medicine to give *on-line* access to its *databases*. Now also used by *Blaise* and a variety of smaller information suppliers. Elhill offers a search language similar to ORBIT (see *on-line searching*).

Em a unit of measurement in printing corresponding to the width of a *lower case* 'm'. It is equal to two *Ens*.

EM 1. end of medium. A *character* in a *string* of *machine-readable data* which indicates the end of the medium on which the data are being recorded. 2. *Excerpta Medica*.

embossment in *optical character recognition*, a measure of the distance between the non-deformed part of a document surface, and a specified point on the printed character.

EMIS Electronic Materials Information Services. A *database* on materials properties and materials supply. Published by *INSPEC*.

EMMS *Electronic Mail and Message Systems*.

EMOL Excerpta Medica On-line: a medical *on-line search* service (see *Excerpta Medica*).

EMS *Electronic Message System*.

emulator *hardware* or *software* which makes a system appear, to other hardware or software, as another system, eg a *word processor* may be able to emulate a *telex*, or a computer of one type may be able to appear to software as a different type of computer.

emulsion in general, a colloidal suspension of one liquid in another. In *micrographics*, it refers to a layer of light-sensitive chemicals in a very finely divided state held in a suspension of gelatine. The emulsion is supported on a base of paper film, glass or plastic to make a *microform*.

En a unit of measurement in printing, corresponding to half the width of a corresponding *Em*, but the same height.

encode to use a *code* to represent *characters*, or groups of characters.

encryption the *coding* of data to protect its privacy, particularly when transmitted over telecommunications links.

end-point determination the process in *speech recognition* for determining the beginnings and ends of words.

end-user a term in information technology to describe the 'final' user or consumer of information.

Energyline a US-based *database* covering all aspects (engineering, economic, social and political) of energy exploitation. Accessible via *ESA/IRS*, *Lockheed* and *SDC*.

Enviroline a US-based *database* covering all aspects of environmental science, ecology, pollution, etc. It is accessible via *ESA/IRS*, *Lockheed* and *SDC*.

EOF a computer statement indicating the end of a file.

EOJ a computer statement indicating the end of a job.

EOM end of message indication when transmitting a signal.

EOR 1. end of record. 2. end of run.

EOT indicates the end of transmission.

EPB Environmental Periodicals Bibliography: a US-based *database* covering all aspects of environment sciences. It is accessible via *Lockheed*.

EPIC Exchange Price Indicators: the *database* of the London Stock Exchange. It incorporates information from international agencies, eg Reuters, Extel, with information from its own sources: the price reporters who tour the floor of the market, and staff who compress business news. *Epic* is made available through the *Topic* information system.

epitome a precise summary of a document.

EPO *European Patent Office.*

EPROM erasable programmable read-only memory. *PROM* which can be erased for reprogramming.

equalization in telecommunications, compensating for *distortion* introduced during the transmission of data.

erasable storage any *storage* medium which can be reused, normally by recording over previous entries, eg an audio-cassette.

ergonomics the study of human-machine interactions.

ERIC Educational Resources Information Center: a *database* covering education and educational resources. It is split into two

main files: a. Current Index to Journals in Education (CIJE); b. Resources in Education (RIE). The database is accessible via *BRS*, *Lockheed* and *SDC*.

Erlang a measure of the traffic on telecommunication circuits. It is obtained by multiplying the number of calls which the circuits carry in one hour by the average duration in minutes of the calls, and dividing the product by sixty.

error a *status word* indicating that the computer has detected an error, and awaits a correction.

error control any system capable of detecting errors and (in the more advanced systems) of correcting them.

error correcting code a *code* that assists in the restoration of a word that has been mutilated in storage or transmission.

error detecting code in telecommunications, a *code* in which each signal is constructed on the basis of a set of rules so that any departures can be detected.

ESA *European Space Agency.*

ESA-IRS European Space Agency – Information Retrieval Service: a *host* providing *on-line* access to some 20 *databases*, mainly scientific, from its headquarters in Italy.

Esanet the *network* providing access to *ESA-IRS*.

ESC escape *character*: a character in a computer data *string* which leads to an exit from a *code*.

escape code *code* used with text input to indicate that the following *character* (or characters) will represent a *function code*.

ESS *Electronic Switching System.*

ESI externally specified index: a feature which enables a computer system to become a central message *switching* centre for a variety of remote devices. It provides automatic routing of messages to and from a main store without disturbing the *program* sequence of the *central processor*.

ETB end of transmission block. Indicates the end of transmitting a *block* of *data*.

Ethernet *network* developed by Rank Xerox to facilitate communication between electronic office equipment (*computers*, *word processors*, complete *work stations*, etc), either locally or internationally.

ETIS-MARFO European and Technical Information Service in *Machine Readable* Form.

ETX end of (transmission) text. An *ETB* message used when the *data* being transmitted represents text.

EURIPA the European Information Providers' Association was formed in 1980 to promote and provide a forum for the European information industry.

EURODICAUTOM European Automated Dictionary. A *pure MAT* (*machine-aided translation*) system, operated by the European Commission, which uses a *terminology bank* to assist translation of scientific and technical documents, in particular those relating to steel manufacture.

Eurolex an on-line search service giving access to case law, legislation and treaties in the UK and Europe (see *on-line searching*).

Euronet a European *packet switching* network for the transmission of *digital* information. It has been established by the European Commission, with entry points in each of the member states. It aims to offer fast, reliable data transmission at an appreciably lower price than existing international tariffs allow (pricing is not dependent on distance). The information search and retrieval services offered through the Euronet system are called *DIANE*.

European Communications Satellite a *communications satellite* program(me), run by *ESA*, which is intended to provide Europe (and possibly other countries) with a satellite communications system similar to those under development in the US.

(See also *Orbital Test Satellite*.)

European Patent Office the EPO, based in Munich and the Hague, has created a central *database* of patents which can be accessed in member states of the EEC via either *Euronet*, or the public telephone network.

European Space Agency a West European organization involved in communication technology as: a. a *database host* (see *ESA-IRS*); b. a launcher and operator of *communication satellites*.

EUSIDIC European Association of Scientific Information Dissemination Centres: an association formed by information suppliers to promote applied technology of information storage and retrieval, as it relates to large *databases*.

EUSIDIC Guide a European guide to *databases* which are accessible *on-line* in Europe. It includes those which are made available via non-European *hosts*, such as *Lockheed* and *SDC*. The title comes from the name of its sponsor (see *EUSIDIC*).

Eutelsat an organization formed by the European *PTTs* to coordinate the interests of the main users of European *communications satellites* (such as *OTS* and *ECS*).

event any occurrence which influences the content of a *data file*.

Excerpta Medica a large medical *database* compiled in the Netherlands and accessible via *Lockheed*.

exchange a unit, normally established by a *common carrier*, to control the passage of communications in a particular geographical area.

execution the performance of operations listed in a computer *program*.

executive system sometimes used as a synonym for *operating system*.

exhaustivity an *indexing* term. It measures how completely the concepts within a document have been indexed. The greater the proportion of concepts covered in the index, the greater the exhaustivity.

expandor an electronic device which expands the volume range of a signal. Used in a *compander*.

experience curve a synonym for *learning curve*.

Experimental Communications Satellite a series of Japanese *communications satellites*. (See also *Broadcast Satellite Experiment*.)

expert system a particular development of *artificial intelligence*. It combines the storage capacity of a computer for specialized knowledge with its ability to mimic the thought processes of a human expert. The *program* for the latter follows a similar learning pattern to a human being's. The computer is provided with a general set of rules instructing it how to reason and draw conclusions. On the basis of these, it

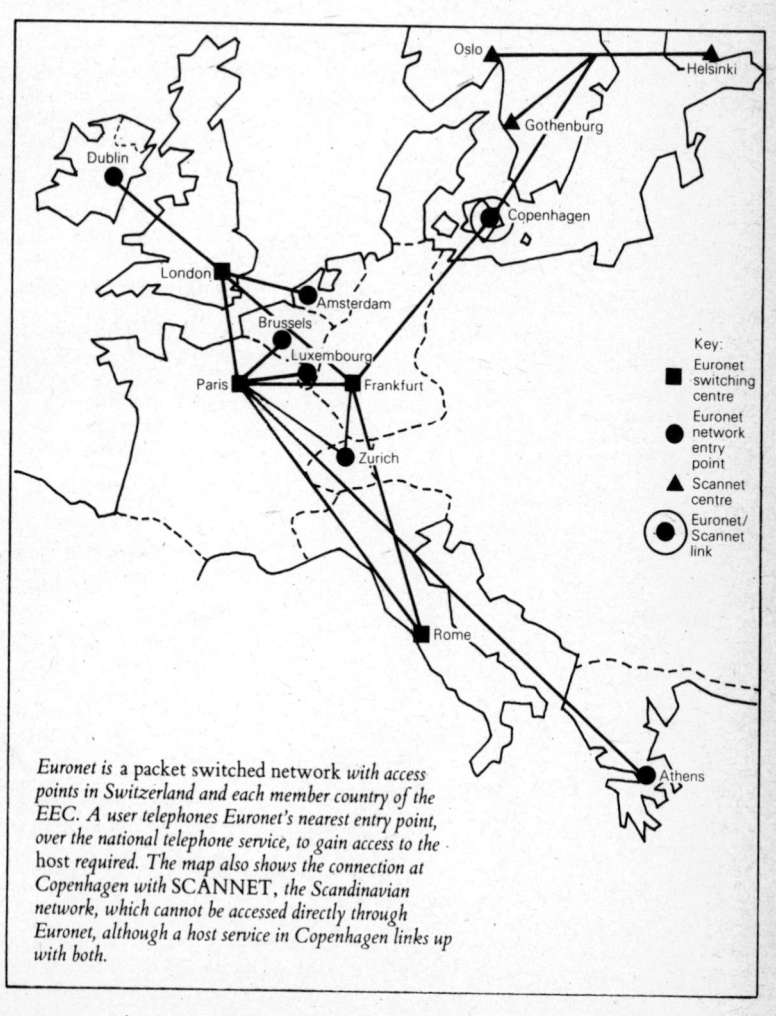

Key:
■ Euronet switching centre
● Euronet network entry point
▲ Scannet centre
◉ Euronet/Scannet link

Euronet is a packet switched network with access points in Switzerland and each member country of the EEC. A user telephones Euronet's nearest entry point, over the national telephone service, to gain access to the host required. The map also shows the connection at Copenhagen with SCANNET, the Scandinavian network, which cannot be accessed directly through Euronet, although a host service in Copenhagen links up with both.

decides which items of information are needed, and continues requesting these, until a conclusion can be reached. In building up an expert system, close interaction is necessary between the computer and the human expert, since the latter's judgement is often based on subconscious lines of reasoning which may be difficult to elicit. Several expert systems are already in use: the number is growing rapidly as computing costs fall. Two areas where expert systems have been operating for some time are:

i. medical diagnosis. The common pattern here is that the system is provided with a range of symptoms, and is taught how to diagnose diseases from them. An extension of this, under development, is to allow a computer to interrogate a patient directly concerning symptoms using either *keyboard* input, or *direct voice input*.

ii. geological prospecting. Here the computer compares the geographical and geological characteristics of an area with its memory of the corresponding characteristics in areas where minerals have been found. Particular effort is being put into the development of expert systems for the detection of oil deposits.

Since expert systems are typically used by people who are not computer specialists, they need to employ sophisticated *high level programming languages*.

extension memory *memory* contained in *external storage* (see *backing storage*).

external memory synonymous with *extension memory*.

external storage see *backing storage*.

extraction indexing the most common form of *automatic indexing*. A document in *machine readable form* is scanned by a computer and words are extracted which comply with a prescribed formula written into an extraction *program*. This program normally directs the computer to extract those words or phrases which occur most frequently, while applying a *stop list* to eliminate common non-substantive words. Less common approaches use relative, as opposed to absolute frequency, as the extraction criterion, and some programs include allowance for word positions and even typeface, eg giving more weight to boldface or italics.

face a particular style of *character*, eg in *optical character recognition* (OCR). (See also *type face*.)

facsimile see *facsimile transmission*.

facsimile laser platemaker a device which uses *facsimile transmission* to transmit a complete page, eg of a newspaper. The transmitted image is used to make a printing plate directly, eg for use in *remote printing locations*.

facsimile transmission a system which can transmit a representation of the form and content of documents over a telecommunications link. It is distinguished from other messages systems in that the recipient receives a complete copy of the original document, not just its information content. It is distinguished from most video systems in that it is concerned with static, not moving images (see *still video*). Current advances in facsimile transmission (fax) techniques suggest that it will be an important method for *document delivery* in the coming decades. Historically, fax has been an *analog(ue)* system, and most current machines are still of this form; but *digital* machines, which can transmit more rapidly, are becoming commoner.
Most commercial fax systems use an electro-mechanical *scanning* technique to convert the tonal variations of the document to be transmitted (the 'subject' copy) into an analog(ue) electrical signal. Scanning can be performed by moving the document, the scanner or both. This causes a spot of light from the scanner to traverse the whole document. Most systems use the reflected light to produce the analog(ue) signal. Depending on the equipment, the document to be transmitted may be wrapped around a cylinder (*cylinder scanning*), or left flat (*flat-bed scanning*).
The diagram illustrates some of the basic differences between facsimile (whether analog(ue) or digital) transmission and conventional digital communication of data. First, the facsimile system has to scan a lot of white space to transmit the letter V, whereas a digital system can put the letter V into binary code, and transmit it as only a few pulses. For transmitting the content of textual information, traditional facsimile

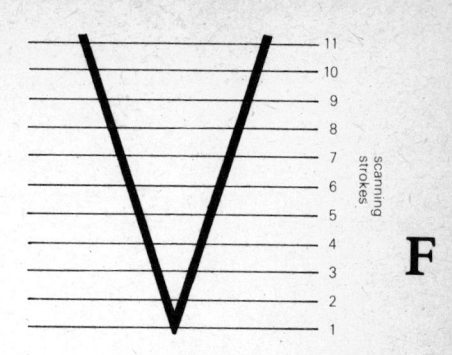

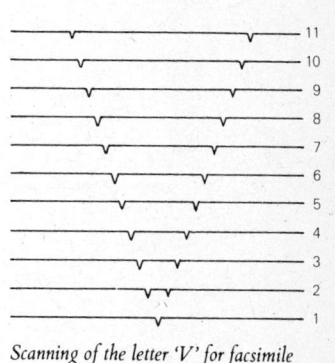

Scanning of the letter 'V' for facsimile transmission (top) and the resulting output (bottom).

F

transmission has therefore often appeared slow and cumbersome. To some extent this is compensated by the fact that, if an error is made in the transmission process, it is highly probable that the letter V will remain recognizable. In a digital system, a single error could result in the printing of an entirely different character.
Facsimile transmission has been available for many decades, and has found some important applications, eg for newspaper printing. However, the development of digital facsimile has opened up new possibilities, making it a potentially important part of new information technology. In a digital facsimile, the analog(ue) information is converted to digital form. Using *data compression* techniques, this can be sent over telecommunications links much faster than is possible in the conventional analog(ue) system.

After transmission and reception, the image must be recorded. Most fax recorders are again either flat-bed, or cylindrical. The main process may depend on electrolysis, electrical resistance, pressure, electro-thermal or electrostatic forces, though other types of recording processes, eg photographic, are also used.

Receiving and sending fax terminals must be *compatible*, so standardization is particularly important. The recommendations of the *CCITT* are followed to ensure this. So far there are three main levels of fax equipment, called Group 1, Group 2 and Group 3. Group 1 was the earliest standard, and corresponds to about six minutes to transmit an A4 sheet. Group 2 takes about three minutes per sheet. Group 3 uses digital techniques, and can transmit a sheet in less than a minute. Group 4 facsimile has not yet been standardized, but is expected to be considerably faster than Group 3.

FACT 1. Fast Access Current Text Bank. An experimental electronic library (at the University of Missouri, US). It produces medical documentation in *facsimile* copy from *microfiche* over a telephone network. 2. Fully Automated Cataloguing Technique. A system used by Library Micrographic Services Inc, US. It combines *computer* control of documents, storage on *microfilm* and *COM*.

FAIRS 1. Federal Aviation Information Retrieval System, US. 2. fully automatic information retrieval system. (See *information retrieval system*.)

false code a *code* producing an *illegal character*.

false drop the retrieval of an unwanted *item* from a *file*, or *database*, as a result of an error in the specification of a request, or *search term*.

false retrieve synonymous with *false drop*.

farad a unit of *capacitance*.

fast access storage 'fast' here is used in a relative sense: whether a storage device is counted as fast depends on the operating speeds of other devices in the system.

For examples of access times, see *storage device*.

fast picture search a feature of more sophisticated *video tape recorders*. It allows the viewer to run quickly through a tape to find a specific point in the recording.

fax abbreviation for *facsimile transmission*, or facsimile communication.

FBR full bibliographic record. The provision for each document of the following information: author, title, place and date of publication, and page numbers.

FCC *Federal Communications Commission*.

FCS facsimile communications system (see *facsimile transmission*).

FD *full duplex*.

FDM frequency division multiplexing (see *multiplexing*).

FDOS *floppy disc* operating system.

FE *format effector*.

feature extraction as the name implies, this covers techniques for extracting significant features from a signal. For example, in *speech recognition*, it refers to methods for determining the *amplitude* spectrum of the incoming speech signal. These transform the *spectrum* into input recognizable to a computer. (Also called pre-processing.)

Federal Communications Commission (FCC) US Government regulatory body for telecommunications.

FEDLINK Federal Library and Information Network. A US *computer network* facilitating *on-line* catalog(u)ing.

FEDREG Federal Register. A *database* covering US Federal Government regulations, proposed rules and legal notices. It is accessible via *SDC*.

feed 1. the process of entering data into a computer. 2. the part of an *antenna* where a signal originates, or is received.

feedback return of a part of the output from a system to its input in order to control the output to within predetermined limits.

feeder cable a *cable* which branches off from a main trunk line in *cable television* in order to serve a group of users, eg all those along a street.

feeder line see *feeder cable*.

FET field effect *transistor*. (See also *MOS*.)

FF *flip-flop*.

fibreoptic (fiberoptic) cable fibreoptic cables are betwen one and two orders of magnitude smaller in diameter than ordinary *coaxial cables* with the same information-carrying capacity. They are light and crush-resistant, and therefore easy to store and install. They also have advantages in performance. They are electrically isolated, so sparks and short circuits are avoided: they neither radiate signals, nor pick up interference.

fibre (fiber) optics see *optical fibres*.

fiche see *microfiche*.

FID International Federation for Documentation. The initials are taken from the French version of the name.

FID/OM *FID* Committee on Operational Machine Techniques and Systems.

FID/TM *FID* Committee on Theory and Method of Systems, *Cybernetics* and Information Networks.

FID/TMO *FID* Committee on Theory, Methods and Operation of Information Systems and Networks.

field in computing, this refers to a section of the computer record which is designated for the storage of specified information. For example, in a *bibliographic database*, a field might cover the data positions where the dates of publication of each document are recorded. A fixed field has a defined, unvarying length, whereas a variable field

can be assigned different lengths.

file refers, in general, to any organized and structured collection of information. The data in such a collection are organized into *items*, and structured so as to facilitate the type of access required.

file control system a system designed to aid the storage and retrieval of data without restricting the type of *input/output* device (see *file*).

file conversion the process of converting a *file* from one medium to another, or one *format* to another.

file interrogation program a *program* designed to examine the contents of a computer *file*.

file inversion see *inverted file*.

file maintenance control of *files* to ensure that their contents are correct.

file management supervision and optimization of work in progress on a computer.

file management program computer *program* which assigns, or recognizes, labels identifying data *files*, and enables them to be called from *storage* as required.

filemark an identification mark to indicate that the last *record* in a *file* has been reached.

file name a series of *characters* used to identify a *file*. The file name is often composed of a *code* which indicates the nature, ownership and/or status of the file.

file security the means by which access to computer *files* is limited to approved operators only. The implementation of file security usually involves the use of *passwords*. (See also *data security*.)

Filmorex system a system for electronic selection of *microfilm* frames.

film recorder *output* device used to produce *COM* and/or to produce record on film of *computer graphics* displays.

film setting often used as a synonym for *phototypesetting*, though it should only be used where the output is on film rather than paper.

filter a device used in telecommunications. It allows signals of specified frequencies to pass without significant *attenuation*, whereas other frequencies are strongly attenuated. The range of frequencies passed is known as the *band-pass* of the filter (see *spectrum*).

fine index a detailed index to a restricted area of information. Normally used in conjunction with a *gross index*, which covers a broader area in less detail.

FINTEL Financial Times Electronic Publishing. Group of *databases* available for *on-line searching* (some on *viewdata*) covering business information. Produced by the British newspaper 'Financial Times'.

firmware a computer *program* written into a *storage* medium from which it cannot be accidentally erased. Often stored in *read only memory* (ROM) which is designed so that it cannot be overwritten. The term also applies to the electronic devices containing such programs.

FIRST fast interactive retrieval system technology (see *information retrieval system* and *interactive*).

FIU Federation of Information Users, US.

fixed field see *field*.

fixed-head *read-write* heads which are kept stationary.

fixed-head disc a *disc memory* with one *read-write head* for each track.

fixed satellite a method of allocating *frequency bands* for *satellite communication* on an international basis which identifies all the sending and receiving stations. (Contrast with *broadcast satellite*.)

FIZ-Technik Fachinformationszentrum Technik. A Frankfurt-based *host* linking with *Euronet Diane*.

flag additional information added to data (normally to each *item*) in order to characterize, or provide information about, the data (or item). Sometimes called a marker, pointer, sentinel or tag. Can also be an indicator in a *program*, used to test some condition set at an earlier point in the program; or a character indicating that the following *code* does not have the normal meaning.

flat-bed plotter an *output* device for *computer graphics*. It uses a pen on a mechanical arm to draw images in several colours on paper. The paper is held on a flat surface (flat-bed).

flat-bed scanner a *scanner* used in *facsimile transmission*.

flat bed scanning a technique used in *facsimile transmission*.

flexible manufacturing system the addition of *robot*-controlled transport of work from one machine to another. Guidance is provided by *numerical control* machines linked into a *computer numerical control* system.

flexography a *letterpress* rotary printing process using flexible, eg rubber or photopolymer, plates.

flicker a visual sensation produced by rapidly alternating light and dark images, when the frequency of the alteration is too slow to allow persistence of vision to give an impression of continuous illumination (see the diagram attached to *flyback*).

flip-flop a *circuit* which can be in one of two states (and so can represent *binary* logic).

flippy double-sided *floppy disc* (but sometimes used as a synonym for floppy disc).

flippy-floppy see *flippy*.

floating point refers to the position of the decimal point in numbers stored in a computer. For example, instead of storing 111.23 and 1112.3 in this form, they may be stored as 1.1123×10^2 and 1.1123×10^3.

a. Floppy disc in its protective cover.

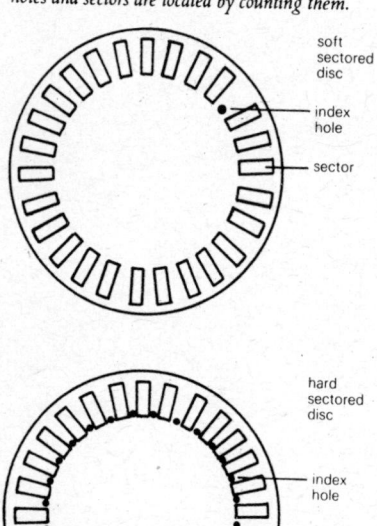

- index hole
- hub opening
- envelope
- read/write head opening

b. Soft and hard sectored discs. Soft sectored discs have only one index hole, the position of the sectors being determined by timing the disc's rotation. Hard sectored discs have many index holes and sectors are located by counting them.

soft sectored disc

- index hole
- sector

hard sectored disc

- index hole
- sector

c. The recorded surface of a disc, illustrating the storage of information in bytes.

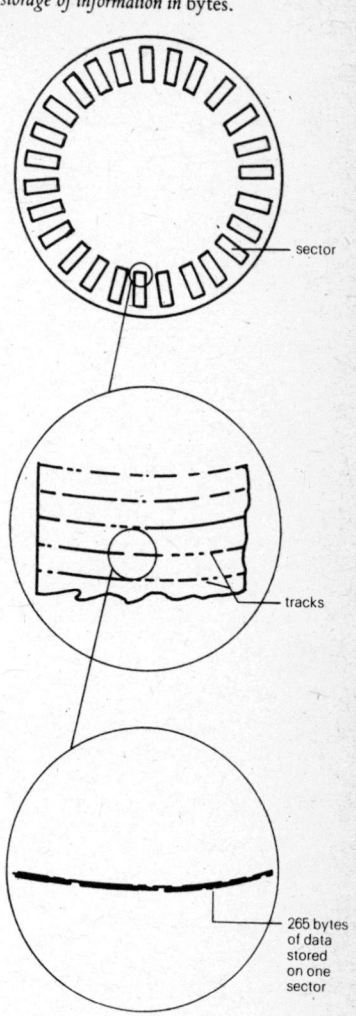

- sector
- tracks
- 265 bytes of data stored on one sector

d. Cross section through a disc.

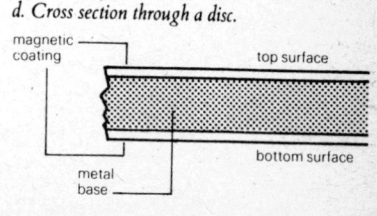

- magnetic coating
- top surface
- metal base
- bottom surface

73

floppy synonymous with *floppy disc*.

floppy disc a disc made of a flexible material, eg plastic, coated with a magnetic surface. Such discs are relatively cheap to make and easy to handle; but they have a more restricted *storage* than a *hard disc*. Floppy discs are usually either 5¼ inches or 8 inches in diameter. See page 73.

flowchart see *flowcharting*, *systems flowchart* and *program flowchart*.

flowcharting a technique for representing a succession of events by means of lines (indicating interconnections) linking symbols (indicating events, or processes). There are two main types – *systems flowcharts* and *program flowcharts*. The first type aims to show the relationship between events in a *data processing* system, while the second aims to break down a problem into the logical components which are analysable by programming commands.

flowchart symbol conventional diagrammatic representation of different events shown on a *flowchart*. The shape of a symbol denotes the type of event (see *systems flowchart* and *program flowchart*).

flow diagram synonymous with *flowchart*.

flow direction the direction of a *flowline* on a *flowchart*. This shows which of two connected events is the antecedent and which the successor.

flowline a line drawn on a *flowchart* to show the relationship between two events.

flush left see *justify*.

flush right see *justify*.

flutter a recurring variation in the speed of a moving medium, eg *magnetic tape*. It is more of a problem for *analog(ue)*, than for *digital* recording. (See also *wow*.)

flyback in a *cathode ray tube* display system, at the completion of the scanning of a line, the beam has to be deflected rapidly to the beginning of a new line. This deflection is

the line flyback. When a whole *frame* has been scanned, the beam is deflected back to the top of the screen for a new frame. This deflection is the frame flyback.

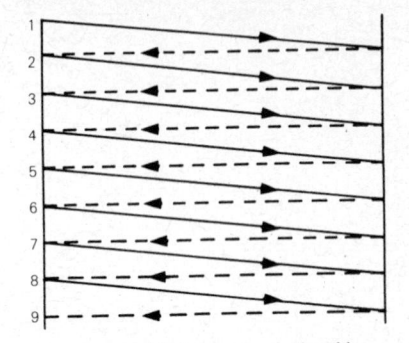

Scanning on a cathode ray tube: the dotted line illustrates line flyback.

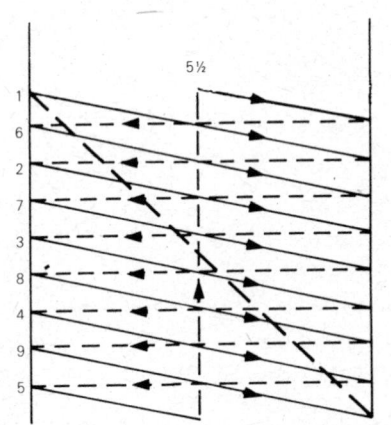

Scanning on a cathode ray tube showing interlacing, which reduces flicker. Dotted lines show line and frame flyback.

flying spot scanning a technique for scanning a surface. A spot, moving at high speed across the face of a *cathode ray tube*, generates light which is then focussed on the document to be read. Used in *OCR* and *facsimile transmission*.

FM *frequency modulation*.

FMS *flexible manufacturing system*.

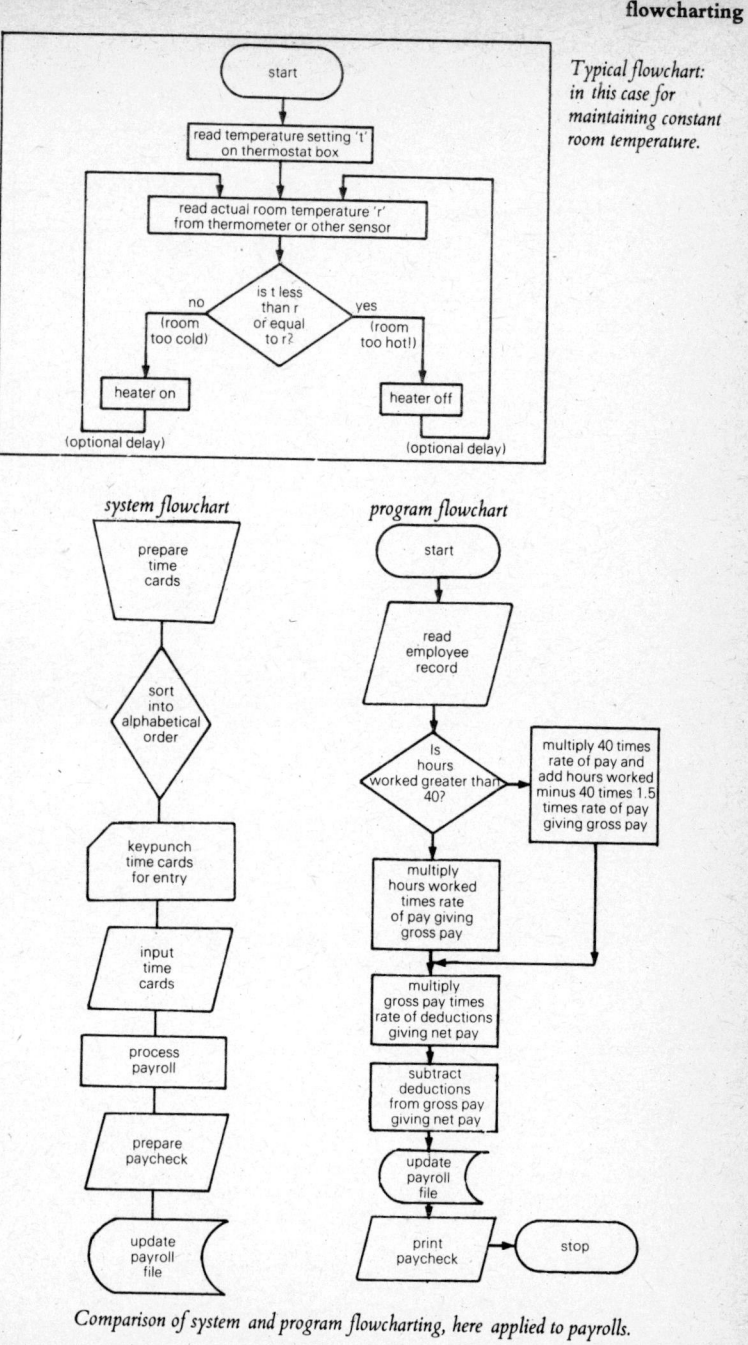

Typical flowchart: in this case for maintaining constant room temperature.

Comparison of system and program flowcharting, here applied to payrolls.

Focus Committee established in 1981 by the UK Department of Industry, to coordinate all British national and international standardization activities relating to information technology.

folio number appearing at the top or bottom of a page.

font US spelling for UK *fount*.

footer information placed at the bottom of a *page*, usually for identification purposes.

footprint the area of the Earth's surface to which a satellite can transmit. A satellite in *geostationary orbit* has a footprint which is about one-third of the surface of the Earth (see satellite communications).

foreground in *time-sharing computers*, refers to high priority tasks which are carried out in preference to those of low priority (*background*). Refers in *word processing* to a computer activity which the operator can identify and use.

foreign exchange service a service connecting a customer's telephone to an *exchange* which does not normally serve the customer's area.

forgiving system a *user-friendly* system which allows inexperienced users to make mistakes without disastrous consequences.

formant human speech covers a range of frequencies. For vowel sounds, the frequency *bands* in which acoustic energy is concentrated are called the formants. Sensitivity to formants is thus important in *speech recognition* and *speech synthesis*.

formant frequencies in speech, the frequencies at which the energy of the acoustic wave peaks. They correspond to the acoustic resonances of the mouth cavity. (See also *speech recognition*.)

formant synthesis see *speech synthesis*.

format a predetermined arrangement of *data*. It may refer, for example, to: the layout of a printed document; the arrangement of *data* in a *file*; the order of *instructions* in a *program*. It can also mean a set of *typographical* commands available at a *keyboard*.

format effector *character*(s) in a *string* of *machine readable data* which are included to determine the layout, or position, of information being transferred to an *output* device.

Forth a *high level programming language* designed to have as many commands as possible in 'plain English'.

FORTRAN an abbreviation for formula translation. A *high level* computer *language*, extensively used for scientific and mathematical programming.

FOSDIC film optical sensing device for input to computers. A device which can be used to *input from microfilm* into a computer *memory*.

fount a complete range of characters, spaces, etc, of one *type* size and design.

fount disc in *phototypesetting*, a glass or plastic disc containing the master *character* images which are used to form typeset characters. Can also refer to master characters stored in *digital* form on a *magnetic disc*.

Fourier analysis an analytical mathematical technique for expressing the waveform of a signal as a sum of related harmonic waves.

fox message a standard message used to test that information is being transmitted satisfactorily. The same message is used to test *keyboard* operation as it includes all the letters in the alphabet. It comes in a number of variants, but a typical form is: 'The quick brown fox jumped over a lazy dog's back'.

fragment codes see *chemical structure retrieval*.

frame this has most currency as a *viewdata* (interactive *videotex*) term. It represents one screenful of information (24 rows of 40 characters each). Several frames may have the same *page* number.

frame flyback see *flyback*.

frame grabber a device installed within a *VDU* which permits the storage and continuous display of a *frame* of information. For example, some *cable television* receivers have a frame grabber which can hold and continually display a single television picture.

framing control *video tape recorder* control which allows a recorder to accept tapes produced on other machines.

free indexing language see *information retrieval techniques*.

frequency see *spectrum*.

frequency division multiplexing see *multiplexing*.

frequency modulation the *modulation* of a *carrier wave* by means of changes to its frequency.

frequency shift keying *frequency modulation* used in the transmission of data.

front end a *terminal* or input device used to create or load data and/or instructions.

front-end system a small *computer* connected to a large (*mainframe*) computer. Used to handle slow *peripherals* for which the power of the mainframe is not required. Relieves the mainframe of tasks which involve delays.

FSK *frequency shift keying*.

FSTA Food Science and Technology Abstracts. A UK-based *database* accessible via *Lockheed* and *SDC*.

FTET full time equivalent terminals. A measure of *terminal* access to a computer system which allows for the number of hours each *terminal* is available for *on-line* access.

FTS Federal Telecommunications System. A network used by Federal Agencies in the US.

full duplex a term used in telecommunications to indicate simultaneous and independent transmission of a signal in both directions along a communication *channel*. (See also *half duplex*.)

function code a *code* which controls the operation of a *typesetter*.

function keys specific *keys* on a *terminal* keyboard which allow the user to issue a series of commands at a single key stroke. These keys can either be designated by the user, or come already programmed in purpose-built terminals. Examples of such terminals are those used in making airline reservations and handling stock market data.

G giga: prefix denoting one thousand million (10^9).

gain the ratio of the strength of an output signal to the strength of the corresponding input signal.

galley proof in hot-metal composition, this is the first *proof*, used for editing in-house and by the author. The document was subsequently produced in *page proof* form. Modern methods of composition normally go to the page proof stage directly, but galleys may sometimes be produced.

gangpunching *punching* information which has been read from a master *card* onto a sequence of other cards.

gap an interval left between *blocks* of data on a *magnetic tape*. It allows the tape to be stopped and started again between *reading* and *writing* processes.

garbage meaningless, or unwanted, data.

gas plasma display a flat panel *display* which can be used as an alternative to a *CRT*. It relies on a dot pattern that can be formed anywhere on the display.

gate the basic building block of *digital* electronics, and therefore of any digital *computer*. A gate recognizes only two possible *input* states. Its *output* state can take only one of these two possible values (see *NAND*, *NOR* for examples of gates).

Gateway *software* which allows users of a *viewdata* (interactive *videotex*) system, such as *Prestel*, access through a viewdata *terminal* to external computers and their *databases*. Data from these computers must be *formatted* into *pages* suitable for viewdata *terminals*. The first software of this type was designed for the West German viewdata system, *Bildschirmtext*.

GDT *graphic display terminal*.

general purpose language a programming *language* whose use is not restricted to a single type of computer, or to a small range of computers. *Basic*, *Cobol* and *Fortran* are examples of such widely applicable languages.

general purpose scientific document writer a computer *output* device capable of producing documents in *hard copy* at the level of complexity required for scientific text (ie including graphs, mathematics, etc).

generation 1. in reference to the devices used in new information technology, the term represents the level of development reached. Thus the earliest computers are referred to as 'first-generation' computers: discussion now centres round the need to produce a fifth-generation computer. (See also *group*.) 2. in reprographic terminology, 'generation' is used to indicate the number of stages required to reach a given point in the reprographic process.

Geoarchive a UK-based *database* covering a wide range of Earth Sciences. It is accessible via *Lockheed*.

Georef a UK-based *database* covering a wide range of Earth Sciences. It is accessible via *SDC*.

geostationary orbit if a satellite is placed in orbit at some 35,700 km above the Earth's equator, it completes one orbit each day. In consequence, since the Earth is rotating at the same rate, the satellite remains stationary above the same point on the equator all the time.
A geostationary orbit cannot be maintained above any part of the Earth except the equator, and, even there, slight perturbations may tend to shift a satellite from its initial position (see *communications satellite*, *satellite communication*).

geosynchronous satellite an artificial satellite in a *geostationary orbit*.

ghost a shadow, or weak additional image, eg on a television screen.

GHz Gigahertz: a frequency of one thousand million hertz (cycles/sec).

GID a West German *on-line* information service which acts as a *host* for a variety of *databases*. The service is available via *Euronet Diane*.

GIDEP Government-Industry Data Exchange Program, US.

giga- prefix denoting one thousand million (10^9).

GIGO garbage in, garbage out – a computing term referring to incorrect *output* resulting from incorrect *input*.

GIRL Generalized Information Retrieval Language. A *search language* developed by the US Defence Nuclear Agency.

global (function) 'global' is used in computer terminology to mean 'complete'. Thus a global search means a complete look through a *file* for a particular item (see, eg *global editing function*).

global editing function a *word processing* activity that interacts with an entire document (ie data *file*).

global search and replace an *editing* function in *word processing*. The user can specify a *character string* in the text and also a replacement for it. The *software* will then automatically substitute the new character string for the old wherever the latter appears in the files. Used for up-dating documents kept in *storage*, eg by substituting a new reference number for an old one in a file of standard letters.

glossary command see *glossary function*.

glossary function in *word processing*, commonly used phrases kept in *storage* which can be inserted at any point in a document by the operator. The phrases are called up by executing a glossary *command*.

golf-ball popular description of the spherical type-head found on the *IBM* electric typewriter.

GPC general purpose computer: as distinct from one constructed for a particular purpose only.

GPO 1. Government Printing Office, US.

2. the former name (General Post Office) of the UK post and telecommunications agency. It has now been split into 'The Post Office' and *British Telecom*.

GPSDW *general purpose scientific document writer*.

GPSS General Purpose Simulation Program: a *high level programming language*.

graph a graphical representation of a relationship by means of dots, lines, curves, etc.

graph follower an *optical scanning* device which reads *data* from a *graph* and translates them into *machine readable* form.

graphic(al) pictorial, or representational, display, eg in new information technology, a computer with a graphics capability. It is to be contrasted with textual presentation (see *graphics*).

graphic arts quality a text which is produced with the *typeface* and range of *characters* normally found in traditional hot-metal *composition*.

graphic data reduction the conversion of *graphic* material into *digital data*.

graphic display terminal 1. a *VDU* which allows the user to display *graphics* material. 2. in *phototypesetting* it has the more specific meaning of a *VDU* which permits the phototypeset matter to be viewed as it appears on the *fount disc*.

graphics *graphic* material produced (in new information technology) as a result of *data processing* (see *computer graphics*).

graphics insertion a *phototypesetting* technique which allows text and *graphics* to be handled together in a single operation.

graphics peripheral an item of *hardware*, connected to a computer and used to *input* or *output* graphics information. Input devices include *graphics tablets* and *light pens*; *output devices* include a wide variety of *visual displays*, *plotters* and *printers*. (See also *computer graphics* and *peripheral unit*.)

graphics plotter a device which provides hard-copy *output* of graphics displayed on a screen, eg a *VDU*, used particularly in conjunction with *computer graphics*. There are two main types: *drum* and *flat-bed*. (See also *printer-plotter*.)

graphics tablet a device for inputting graphics. Using a *stylus*, diagrams, maps, charts or free-hand drawings can be created, and appear instantaneously on a display screen. The tablet can also be used to manipulate the image, or to direct it to a *storage device* for subsequent recall, or transmission.
Tablets are normally made of a grid of wires partially embedded in a thin substrate. When a tablet is touched by its stylus, the wires are brought into electrical contact. This produces a pulse which enables the computer to register the position of the stylus.
A graphics tablet is sometimes referred to as a 'digitizing tablet'. There are types which operate on other principles, eg an *electroacoustic tablet*.
(See also *computer graphics*.)

graphic structure input a method of input for *on-line searching* of a *database* containing details of chemical structures, eg *CAS on-line*. It allows the user to phrase a search query by constructing the required structure on an *intelligent graphics terminal*. Structural features are either selected from a *menu*, or by manipulation of a *cursor* from a *keyboard*. When the required structure has been composed, the *database* is searched for any corresponding chemical structures and details are listed (see *chemical structure retrieval*).

gravure a method of printing from an etched, or sunken, surface (also called 'Intaglio'). It is mainly used for long print runs of illustrated magazines.

gray scale see *grey scale*.

Green Thumb the first federally funded *videotex* system in the US. It is an experimental system supplying information to farmers (prices, weather, etc).

Gremas code a *code* for *chemical structure retrieval*.

grey literature this term is often used for 'semi-published' literature: that is, literature which is not formally listed and priced, but is nevertheless in circulation, eg institutional reports. Such literature is often particularly difficult to trace. Hence, inclusion of grey literature in *databases* available for *on-line searching* greatly improves its bibliographic control, and so its availability (see *SIGLE*).

grey scale a range of different *tones* in a *continuous tone* image. A digitized grey scale may record anything from 8 to 255 or more different density levels.

grid 1. a device used for measuring characters in *optical character recognition*. 2. in *phototypesetting*, a carrier for the master character images which are to be scanned. 3. a device used for text alignment in making film images.

gross index an index with wide coverage, but little fine detail. Having located the appropriate domain in a gross index, a user then searches a *fine index* to locate specific entries.

Group (1, 2, 3, 4) the name used to describe *facsimile transmission* equipment conforming to the recommendations of the *CCITT*, eg Recommendation T2, 'Standardization of Black and White Facsimile Apparatus', covers 'Group 1' systems.

grouping the collation of data which come under the same classificatory heading.

group mark *character* which indicates the beginning, or end, of a set of *data*.

group SDI a *selective dissemination of information* (SDI) service in which group, rather than individual profiles (see *user interest profile*), are set up and matched against additions to a *database*. The database provider defines a set of standard profiles. Users subscribe to one, or more, of these. This approach offers less precision than can be obtained with individually tailored profiles. However, Group SDI is obviously cheaper for the subscriber, since the cost of each *run* is shared.

GSI *graphic structure input*.

GSIS Group for the Standardization of Information Services, US.

GT&E General Telephone & Electronics. A large US *common carrier*.

guard band in telecommunications, an unused band of *frequencies* between two assigned frequencies, which provides a safety margin, or guard, against mutual interference.

guide a device that identifies the tape path in audio or *video recorders*.

gulp a group of *bytes* treated as a unit: similar to a *character* or *word*.

H Henry: *SI* unit of inductance.

HAIC Hetero-Atom in Context. *Indexing system* used for heterocyclic chemical compounds in the *CA database*.

HAL Harwell Automated Loans system. The computerized loan system of the Library, UK Atomic Energy Authority at Harwell.

half duplex in telecommunications, transmission of signals along a communications *channel* in both directions, but not simultaneously. (See also *full duplex*.)

half-tone a print which appears to be continuous, but actually consists of small, closely spaced spots of varying size. Typically used for illustrations in newspapers.

half-word half of a computer *word*. For example, in a computer that works on sixteen-*bit* words, a half-word is eight bits.

Hamming code a method of achieving data integrity (named after its inventor) by detecting and correcting transmission errors. It is a term used in *teletext*.

handshaking the exchange of alerting signals between transmitting and receiving points prior to full transmission between the two.

hard copy normally synonymous with print-on-paper (ie text and graphics recorded on sheets of paper).

hard disc a *magnetic disc*, used for bulk storage of computer data. Hard discs offer much larger storage capacities than *floppy discs*, but have to be handled more carefully. They may need to be operated in a clean (filtered air) environment. *Main-frame* computers typically employ hard discs. The alternative name 'rigid disc' is sometimes used.

hard sectoring the physical marking of *sector* boundaries on a *magnetic disc* by punching holes in the disc. Contrast with *soft sectoring*. In hard sectoring, all available space can be used for data *storage*.

hardware the mechanical, magnetic, electronic and electrical devices which go to make up a computer.

hardwiring permanently wired electronic components capable of *logical decisions*. *Intelligent terminals* operating without *software* are hardwired. The *program* logic cannot be changed in a hard-wired computer (except by replacing the circuit boards or memories). Not to be confused with a *dedicated* computer, which is a general purpose computer assigned for a specific task.

Hartley a unit of information content. It is equal to one decadel decision (ie the designation of one of ten possible, and equally likely, values or states).

HD 1. *half duplex*. (Also abbreviated to HDX.) 2. *high density*.

HDX *half duplex*. (Also HD.)

head a device which reads, records or erases information on a *storage* medium.

head drum the part of a *video recorder* which contains the rotating *heads*.

header 1. information placed at the top of a *page*, usually for identification purposes. 2. the initial part of a message which contains the information necessary to direct the message to its destination(s).

header file in *information retrieval systems*, the *file* containing the complete records of a particular *database*, usually in order of accession.

header sheet an instruction sheet for an *OCR* device which informs it of the *format*, etc, to be expected on subsequent sheets.

heading card a card containing significant information which is used in printing headings, eg relating to *index terms*.

head rotor a rotating drum with one or more recording, or reading, heads mounted on its periphery. It is used in *video tape recorders*.

helical scanning a technique by which the *head* in many *video recorders* reads information from a *tape*. One or more heads are mounted on a drum which rotates at high speed. The tape is wound on the drum helically (so requiring a corresponding *scanning* pattern). Although the tape moves at a relatively slow speed, the head-to-tape speed can be quite high.

helical waveguide see *waveguide*.

Hermes a UK experimental electronic *document delivery system* being developed by *PIRA* and the UK Department of Industry. It will link terminals to a *database* dealing with current affairs.

Hertz *SI unit* of frequency (= one cycle/second).

heuristic solving a problem by means of trial and error.

hexadecimal code a data *code* which uses the base 16 (as compared with a base 2 for a *binary* code, and ten for a decimal code).

hexadecimal keyboard *keyboard* with 16 *keys*: 0-9, plus A,B,C,D,E and F to represent 10-15. These keyboards (also called *hex pads*) are often used in conjunction with *microprocessors* (see *hexadecimal code*).

hex pad see *hexadecimal keyboard*.

HICLASS *hierarchical classification*.

hierarchical classification a framework for designation in which terms are arranged according to a hierarchical principle: ie the classification splits items into initial sets, and then successively splits these sets into ever finer sets. (See, for example, *decimal classification*.)

high density the provision of a relatively high *storage capacity* per unit storage space, eg in *bits* per inch.

high level programming language a computer language which allows users to employ a notation with which they are already familiar, eg it may include such terms as: if, then, print, +, etc. Each

natural language instruction actually corresponds to several *machine code* instructions. The most common high level languages are *ALGOL, BASIC, COBOL, FORTRAN* and *PASCAL*.

high performance equipment equipment producing signals of a quality suitable for transmission via *telephone* and *teleprinter* circuits.

high speed printer a computer *printer* which can operate sufficiently rapidly to be compatible with *on-line* printing.

high speed reader an *input device* capable of obtaining data very rapidly from an input medium, eg *card* or *tape*, or from a *storage* device. High speed card readers operate at 1000 cards per minute, or more. High speed *punched paper* tape readers operate at 500 *characters* per second, or more.

HISAM Hierarchical Indexed Sequential Access Method. This refers to a method of organizing a *database* for *disc storage*.

Historical Abstracts a *bibliographic database* covering world history from 1450 to the present. It is accessible via *Lockheed*.

hit in *information retrieval systems*, a hit occurs when an enquiry is successfully matched with a record in the *database*: ie a *search term* is matched with an *index term* or *keyword*.

hit-on-the-fly printer a *printer* where the type does not stop moving during the impression time. The need to stop and start is avoided, so saving time and wear.

HMSO Her Majesty's Stationery Office: the UK Government publishing office.

holding time the length of time a communication *channel* is in use for an individual transmission.

Hollerith code a standard code for *card punching*.

hologram see *holography*.

holographic storage holograms are currently being developed as high capacity information *storage devices*. A 4 x 6-inch *microfiche* could hold as much as 200 million *bits* of information in 20,000 holograms (see *holography*).

holography the creation of three-dimensional images of objects using light produced by *lasers*. A typical arrangement for producing one type of 'hologram' (also called a 'holograph') is shown in the diagram.

The beam of a laser is split into two. One part is reflected onto a photographic plate. The other is directed onto the object (in this case, a transparency) and the light transmitted to the same photographic plate. The two beams produce an interference pattern on the exposed plate which is then developed to produce the hologram. If the hologram is now placed in the path of the reference laser beam, a reconstructed image of the original object is produced.

In the longer term, it may be possible to use moving holographic images as an educational, or entertainment, medium, eg 'holographic television'. For communications, 'holophones' have been proposed, where speakers at distant locations can see 3-dimensional images of each other.

In the shorter term, holographic techniques may find application in information storage (see *holographic storage*).

home the starting point for a *cursor* on the screen of a *VDT*.

home information system a system which allows the home to be used as a centre for electronic control and communication. It is normally envisaged under this heading that activities such as domestic accounting, education, energy management, shopping and banking will be carried out from the home using equipment based on the *chip*. See diagram overleaf.

horizontal raster count the number of horizontal divisions in a *raster*. (See also *raster count*.)

host (sometimes also called an information spinner, information vendor, or on-line retailer) a host is an entrepreneur who makes available a number of *databases* through his own computer. Users are charged for access to these databases (usually via *on-line searching*). Conversely, the database compilers are paid either a commission or a fixed rental. Host computer is also more generally used to refer to a primary or controlling computer in a multi-computer system.

hot-metal (composition) mechanical *composition* in which metal *type* is cast anew for each job, and then melted down for re-use. Monotype is a typesetting device which casts single type characters. A line-caster, either an Intertype or Linotype, casts one complete line at a time.

hot spot a bright spot in the centre of a projected image. A term used in the display of *microform*.

house corrections the corrections of

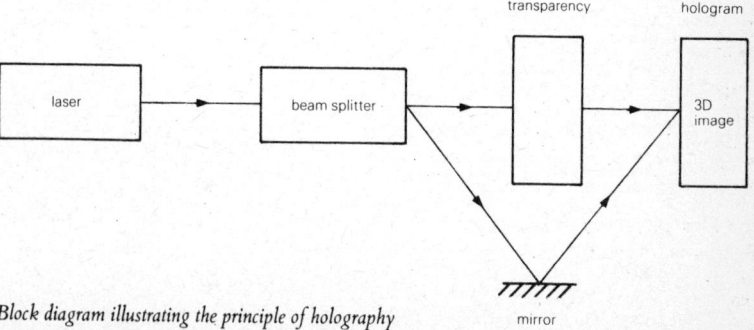

Block diagram illustrating the principle of holography

errors introduced at the printers either by an operator at the *keyboard*, or by a malfunction of the *typesetting* equipment.

housekeeping routine maintenance of *programs* and other contents of a computer.

HSELINE a *bibliographic database* produced by the Health and Safety Executive in the UK.

HSM high speed memory. Memory which has very short *access time* (see *RAM* and *storage devices*).

HSP *high speed printer*.

HSR *high speed reader*.

human-aided machine translation (HAMT) a *machine translation* system in which human intervention is only needed to resolve semantic or syntactic ambiguities, and problems arising from non-literal usage. The computer carries out the basic processing of *source* and *target language* texts. The major advantage of HAMT systems is that they do not require truly bilingual operators. Monolingual operators with some competence in both *source* and *target languages* are sufficient. The main disadvantage is that HAMT systems are only practical (in terms of cost-effectiveness) with *computer algorithms* powerful enough to require infrequent human clarification. Otherwise, HAMT systems cannot compete with conventional and *machine-aided translation (pure MAT)* systems (see *junction grammar* and *Mind*).

hybrid computer a *computer* system which combines *digital* and *analog(ue)* computers.

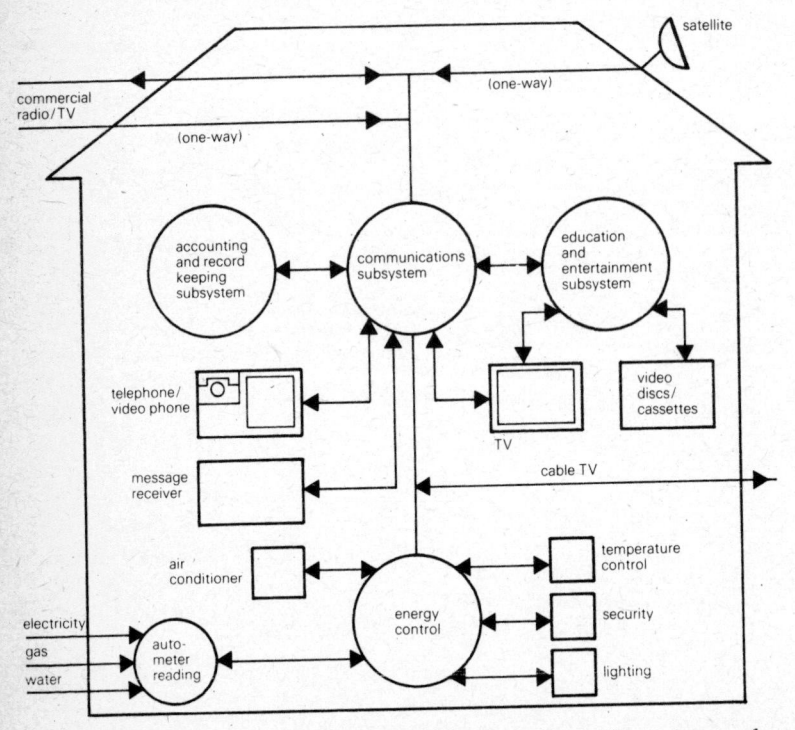

Integrated information system such as can be incorporated into a domestic dwelling, making use of a wide range of information technology inputs.

Its main advantage is the combination of a digital computer's *storage* facilities with an analog(ue) computer's speed of *data* integration.

hybrid computer system a system which combines *analog(ue)* and *digital* devices.

hybrid interface an interface between a *digital* and an *analog(ue)* device.

hyphenation breaking a word at the end of a line, so that the line can be both left- and right-*justified*.

Hz *Hertz*

IAA International Aerospace Abstracts (see *NASA*).

IACBDT International Advisory Committee on Bibliography, Documentation and Terminology, UNESCO.

IACDT International Advisory Committee for Documentation and Technology, UNESCO.

IAD initiation area discriminator: a type of *cathode ray tube*.

IADIS Irish Association for Documentation and Information Services.

I & A indexing and abstracting (see *indexing* and *abstract*).

IARD Information Analysis and Retrieval Division of the American Institute of Physics.

IBA Independent Broadcasting Authority, UK.

IBIS International Book Information Service. A UK *current awareness service*. It mails information on recently published books to subscribers, whose areas of interest are held on computer files. It is also an *information provider* on *Prestel*.

IBM International Business Machines: the world's largest computer manufacturer.

ICIC International Copyright Information Centre, UNESCO.

ICIREPAT International Cooperation in Information Retrieval among Examining Patent Offices. Based in Geneva, its objective is to promote international cooperation in the documentation and retrieval systems operated by national patent offices. Also known as CICREPATO.

ICL International Computer Laboratories: Britain's largest computer manufacturer.

ICR International Council for Reprography.

ICSSD International Committee for Social Sciences Documentation and Information.

I-cycle see *machine cycle*.

ID identification. A unique set of *characters* which are assigned to a computer user and employed during a *log-on*.

IDC Internationale Dokumentationgesellschaft für Chemie. A joint venture of twelve chemical companies in West Germany, Austria and the Netherlands. Its aim is to develop and operate a computerized *information storage* and *retrieval* system for chemical information.

IDD International *direct distance dialling*.

IDDS International Digital Data Service. A *data transmission* service.

identification the coded name assigned to an item of *data*. It normally consists of a series of *characters*.

identify to assign a *label*, either to a *file* (ie to create a *file name*), or to *data*.

idle time time during which a machine is ready for operation, but is not in actual use.

IEC *International Electro-technical Commission*.

IEEE Institute of Electrical and Electronics Engineering, US.

IERE Institution of Electronic and Radio Engineers, UK.

IFD International Federation for Documentation: also known as FID.

IFIP International Federation for Information Processing.

IFIPS International Federation of Information Processing Societies.

IFRM see *International Frequency Registration Board*.

IIC *International Institute of Communications*.

I

I Inf Sc Institute of Information Scientists: a UK professional body, also known as *IIS*.

IIS see *I Inf Sc*.

ILL *inter-library loan*.

illegal character a combination of *bits* (a *false code*) which a computer is unable to recognize as representing a *character*.

image data tablet see *data tablet*.

image dissector 1. a mechanical, or electrical, device which detects the light level in *optical character recognition*. 2. a detector for scanning the image produced by a photo-cathode.

image printer a *printer* which uses optical technology to compose the image of a complete page from *digital input*. The final copy is usually produced as print on paper. Unlike an *intelligent copier*, an image printer cannot produce prints directly from *hard copy*.

image processing the processing of images using computer techniques. This can cover a variety of processes, including enhancement of images, extraction of particular features, *digital* storage of images for transmission or later retrieval, etc.

IMDS International Microform Distribution Service (see *microform*).

impact printing a conventional means of printing in which a hard die hammers an inked ribbon onto the paper. It is used, for example, in typewriters, many dot *matrix printers* and *line printers*.

imposition a printing term. The placing of several pages together on a sheet with correct margins, so that, when the sheet is folded, the pages will appear printed in the correct sequence and position.

IMRADS information management, retrieval and dissemination system (see *information retrieval system*).

incompatibility see *compatibility*.

indent additional space inserted before the first word in a line.

index at the most general level, an index consists of a series of identifiers each of which characterizes a document, *abstract*, or other piece of information. These identifiers can be arranged in a variety of ways to suit user needs. Examples are indexes of authors, titles, dates, countries, institutions, report numbers.
The most complex indexing operation is usually the production of subject indexes. The type of indexing system used is often called the index language. (See also *information retrieval system* and *indexing*.)

indexing the process by which 'labels' are produced for documents, or for information. These labels are used for subsequent retrieval of the original document, or information (see *information retrieval techniques*). Many of the items contained in a document can be used to provide indexing terms, eg author, author affiliation, subject, providing a wide variety of *indexes*. In derived-term indexing, the indexing terms are taken from the document itself. In assigned-term indexing, the indexer assesses the document and decides what terms to apply to it. Thus *natural language* indexing is 'derived term'; whereas indexing using a list of subject-headings is 'assigned term'.
In pre-coordinate indexing, the indexer combines more than one term to describe a document, and the document can be found listed under the combination of terms. In post-coordinate indexing, the indexing terms are assigned individually, and the searcher uses his/her own combination of terms. Most advanced computerized information systems, particularly *bibliographic databases*, are post-coordinate. Classification systems can be regarded as pre-coordinate indexing systems, where the terms are arranged according to their subject relationships. (See also *automatic indexing*.)

index register a device for automatically changing *addresses* of computer *instructions* whilst still storing the instructions in the *memory*.

index term a term used to classify a document, or item, in a *database* (see *keyword*, *information retrieval system* and *techniques*).

indicative abstract see *abstract*.

INFCO Information Committee of the *International Standards Organization*.

inferior a synonym for *subscript*.

INFOL Information Orientated Language. A *high level programming language* developed by the Control Data Corporation, US.

INFORM also called ABI/INFORM. A US *database* containing business information. It is accessible via *BRS*, *Lockheed* and *SDC*.

informatics a word derived from the French 'informatique'. It covers the study of information and its handling, especially by means of *new information technology*.

information *data* processed and assembled into a meaningful form.

information abstract see *abstract*.

information channel the *hardware* linking two *terminals* in a data transmission link.

information feedback *data* received by a *terminal* are retransmitted to the sending terminal for checking.

information provider this term is used mainly in the context of *viewdata* (interactive *videotex*) systems. It describes an individual or organization providing material for the *database(s)* involved.

information retrieval system an information retrieval system basically provides information to users in response to their requests. In information technology, the term refers to the methods and processes whereby computer *files* containing information can be searched for particular *items*, in response to the definition of particular user needs. Since the performance of an information retrieval system is necessarily related to the ways in which information is stored and categorized, the expression 'information retrieval system' is normally used to include *data* and file compilation and *storage*, in addition to the search for, and delivery of, information items. More specifically, the performance of a retrieval system depends on the ability to maintain a consistent and clear boundary definition to its coverage, to gather information within that boundary, and to develop *indexing systems* which provide the closest possible coincidence between the classification of items within the *database* and the description of their information needs by users (see *completeness*, *relevance*). This initial classification of additions to the database is sometimes carried out manually, by allocating multiple indexing terms (often called *keywords*) to each incoming item. Alternatively, and with increasing frequency, automatic indexing methods are used. These have greatest applicability for *bibliographic databases*, where a document's content can be indexed by means of *keyword-in-context* terms. These are allocated by the computer on the basis of an analysis of individual words, and groups of words, in document titles. Items once indexed, are entered on files. Several different types of file organization are commonly used, the simplest of which is the *serial file*: a sequential listing of items with no indication of the links between them. While economical in storage space, a sequential search of a complete database can require a great deal of computer time. This problem is most frequently overcome by creating *inverted files*: an inverted directory is used to store, for each applicable keyword (or indexing item) the corresponding set of document or item identifications and locations.

Files can be held on a variety of different storage media; the form chosen depending on the type of file organization adopted and the amount of data contained. *Magnetic tape* is usually used for *serial files*, since the whole file has to be searched to meet each user request. If, however, index terms can direct the search to particular items, eg via inverted files, then *direct access storage devices*, eg *magnetic drums* and *magnetic discs*, can provide a far more rapid response. The actual search of a database can be carried out either *off-line* or *on-line*. Off-line systems have cost advantages and have proved useful in servicing *selective dissemi-*

nation of information systems and other current awareness services. They also feed into many *secondary services*. *On-line searching* has the advantage of being *interactive*, permitting 'browsing', and giving immediate results. Information retrieval systems are thus increasingly going on-line (see *on-line searching*).

information retrieval techniques most existing computer information retrieval (*IR*) systems are actually systems for the retrieval of bibliographic references to articles, books, reports, patents, etc. But there are IR systems for retrieving numerical and business data, and, with the advent of *videotex*, a wide range of information of all kinds if available via computer. Various techniques can be used to extract information from these systems. Their characteristics can be described in terms of bibliographic retrieval.

There are many ways of characterizing a document, eg by author name, or subject: these are the identifying 'labels' of the document. The essence of most IR is the matching of these 'labels' with enquiry 'labels' used by the person who is interrogating the system. When the labels match, there is said to be a 'hit'; and the user can then retrieve the bibliographic references to the matching documents. Some very sophisticated systems have been developed on the basis of this relatively simple concept.

The documentary labels typically employed for retrieval include: name of author, or authors; an article/book title; a number, eg *ISBN*, patent number, report number; a *Copyright Clearance Center* number; a publisher's name; a place of publication; a volume number and a part number (for *serials*); page numbers; a date of publication; a date of submission and a date of acceptance, eg for research papers; the form of publication, eg paper or *microform*; author affiliation; the body providing funds for the work; the language of the document; an *abstract* or summary; a list of references or *citations* to other documents or work; contents list, index, glossary, etc. Then there are items from the contents of the document, eg the text itself, data, tables, figures, captions.

In addition there are labels which can be added to the document – entered either in the course of publication, or by an indexer prior to entry of the bibliographic description into the *database*. Examples of these are: a classification code, eg *UDC* and/or a classification specific to the database system; index terms or phrases describing the contents; data or other *flags* indicating the presence of some kind of information, eg numerical data in the contents.

In principle, all these items, both inherent and added, could be used as 'labels' to be matched against enquiries. In practice, no system makes use of them all. Searching a database consists of formulating an enquiry by combining the labels of interest in a way that the computer can understand. This is normally done by *Boolean logic*, which basically consists of AND, OR and NOT. Suppose we wished to find out whether our database includes any documents on land reptiles or freshwater fish (except trout). We would express this as: ((freshwater AND fish) OR (reptiles AND land)) NOT trout. The exact formulation will depend on the structure of the database, and its associated *command language*.

The database may be arranged in a *serial file*, where each *record* represents a document, or it may be in *inverted file* form, where each record contains some indexing term (or label) with a list of the documents to which that label applies. In a serial file search on a particular topic, each item must be searched in sequence; whereas in an inverted file only those records headed by the labels of interest need be accessed and searched. An inverted file usually requires the use of a subsidiary file, which gives a description of each document found in the search.

The terms that can be used in a search expression may be restricted by the nature of the database and by the *indexing language* used. For example, the terms denoting the subject content of a document (often called *descriptors*) may be controlled, so that problems due to synonyms, etc, are less likely to arise. Thus, in an *uncontrolled* (or free, or natural) language system, a search done on the phrase 'heart failure' would not retrieve items that used the expression 'cardiac failure' or 'cardiac arrest'. An essential aid to such bibliographic control is a listing of all the terms to be used, and their relationships. This is called a *thesaurus*.

Controlled language indexing has the advantage of synonym control. It can also help to restrict (or expand) a search in a controlled way, using the known relationships between the indexing terms. It has the disadvantage that the index can become rapidly out of date, and the corresponding danger of not being fully comprehensive.

Some systems use both controlled and natural language indexing, whilst others allow variations of the search terms themselves, eg truncation, or *weighting*.

It can be seen that the retrieval technique has an intimate relationship to the indexing system. There are many such systems which are, or can be, applied to computer retrieval (see *word/character frequency techniques, citation indexing*).

information spinner see *host*.

information storage and retrieval system see *information retrieval system*.

information storage and retrieval techniques see *information retrieval techniques*.

information technology the acquisition, processing, storage and dissemination of vocal, pictorial, textual and numerical information by means of *computers* and *telecommunications*.

information theory theory concerning the measurement of quantities of information, and of the accuracy of information transmission and retrieval.

information transfer module a device (developed by *ITT*) which permits intercommunication (ie provides an *interface*) between *telephone*, *telex* and data *terminals*. ITM converts the signals of each system into a form recognizable by either of the other two.

information vendor see *host*.

INFRAL Information Retrieval Automatic Language. A special computer *language* providing the ability to construct *bibliographies* from indexed information.

infrared electromagnetic waves in the *band* between approximately 0.75 and 1000 micrometers. Certain types of transmission, eg via *fibreoptic cables*, employ these waves.

in-house line a privately owned, or leased, line connected to a public network.

in-house system a communications *network* which is either contained in one set of buildings, or which, at least, does not use *common carrier* facilities.

INIS International Nuclear Information Service. This relies on a *database* covering all aspects of nuclear science and technology (technical, economic, social and political) compiled by the International Atomic Energy Authority in Vienna. *On-line* and *off-line* search services are offered.

initial program load the procedure which causes an *operating system* to commence its operation.

ink jet printing a jet of liquid, issuing from an orifice, will break into droplets, if it is vibrated at an appropriate frequency. A charged electrode is placed near the jet, so that each droplet carries a charge. These charged droplets are then deflected by an electrostatic field. The deflection can be varied so that the ink drops can be directed to particular parts of a paper sheet. Ink-jet printers are controlled by digitally-stored information. A main feature is that each printed copy can differ, since the printing is individually controlled. Diagram overleaf.

ink uniformity in *optical character recognition* (OCR), this refers to the variations in light intensity over the surface of characters.

Inmarsat a marine version of *Intelsat* designed to provide *satellite communications* systems for the merchant navies of the world. The initial talks establishing Inmarsat were held in 1979.

Inpadoc the largest computerized patents *database* in the world, it is estimated to hold 98 per cent of the world's currently published patent documents. The annual number of updates is around one million items. The database is accessible via *Pergamon-Infoline* and *Lockheed*.

INPI-1 *database* produced by the French Patent Office, covering French patent information. It is accessible via *Télésystème*.

INPI-2 *database* produced by the French Patent Ofice, covering European patent information. It is accessible via *Télésystème*.

in-plant system a synonym for *in-house system*.

input information received by a computer, or its *storage* devices, from outside.

input bound a system where speed of performance is restricted by the capability of the *input* system, eg a *phototypesetter* able to work at over 1000 *characters* per second, but limited to 500 per second by a *paper-tape* input.

input device any item of equipment which permits data and instructions to be entered into a computer's central *memory*, eg *MICR*, *OCR*, *keyboards*, *terminals* and *light pens*.

input limited refers to a situation where the input speed is the factor which limits the rate of processing. (See also *input bound*.)

input/output controller a device with an independent *data processing* capability. This offers additional independent paths between the *central processing unit* and its *peripherals*. It thus increases the number of *input/output* channels available to the system and speeds up its operations.

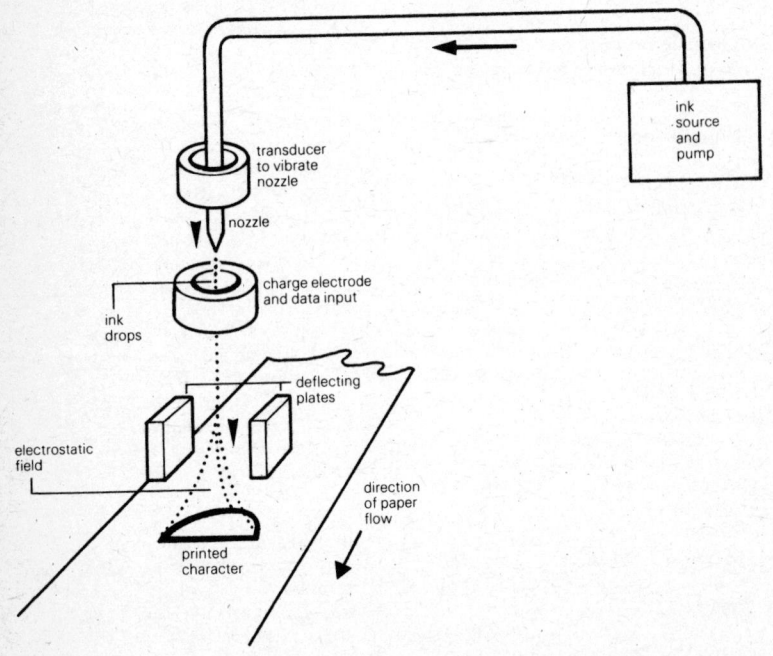

Diagrammatic portrayal of the process of ink jet printing. The jet of ink is directed onto the surface of the paper sheet, to form the required character, by breaking it into droplets under vibration of the nozzle transducer, charging the droplets by passing them close to an electrode and then deflecting them very accurately by an electrostatic field that is directed by the computer data source.

input/output control system the *hardware* and *software* which handle the transfer of *data* between the *main storage* and *external storage devices*.

inscribe the preparation of a document which is to be read by *optical character recognition*. It usually involves recomposition of some of the characters in the document.

INSIS Inter-Institutional Integrated Services Information System. A proposed information system linking the *CEC* to key national ministries and parliaments. The system aims to combine several services: *telephone, telex, teletex, fax, electronic mail, teleconferencing* and *word processing*.

INSPEC Information Services: Physics, Electrical and Electronics, and Computers and Control. Provides databases in these subjects which are made available for *on-line searching* via a variety of *hosts*. INSPEC also offers *magnetic tapes* for sale, and SDI services.

INSTARS information storage and retrieval systems (see *information retrieval system*).

instruction a command (usually in the form of a *character string*) to a computer to carry out some operation.

instruction repertoire see *instruction set*.

instruction set the range of different *instructions* which an operator can use with a particular computer.

instrument in telecommunications, this is often used in the specific sense of a device used to originate and receive signals, eg telephone handsets, computer *terminals*.

Intaglio a printing method from a recessed image. Used for *gravure*, but also refers to copper-plate, die stamping, etc.

integer whole number.

integrated circuit see *chip*.

Integrated Services Digital Network a *network* carrying *digital* information which

usually combines voice and data in the same channels to provide a wide range of communication options.

intelligence 1. describes the *data processing* capability of a computer. 2. the ability to learn and to improve a system in *artificial intelligence*.

intelligent copier any copying device that uses a *microprocessor* to control its functions. Such copiers can often not only produce and collate copies made from print on paper, but can also accept *digital information* as an *input*, and use it directly to produce *hard copy*.

intelligent terminal a *terminal* which can be used to perform local *data processing* without the help of a *central processor*.

Intelpost a US and UK *facsimile transmission* service offered by *Intelsat*.

INTELSAT International Telecommunications Satellite Consortium. The name 'Intelsat' refers both to a *communications satellite* organization and to the satellite that it launches. The organization is currently the largest civilian system in this field. It was established in 1964 by eleven nations, and now has over a hundred member countries. Its first communications satellite – Intelsat I (popularly known as 'Early Bird') – was launched in 1965. The current model – Intelsat V – has a capacity some hundred times that of Intelsat I, in terms of channels available. Several are kept in *geostationary orbit* simultaneously, and handle both telephone and television channels. The different nations' share in Intelsat depends on their proportionate use of the system. In consequence, the US has been the leading shareholder to date (via *Comsat*), followed by the UK (see *satellite communication*).

interactional (interactive) mode synonymous with *conversational mode*.

interactive the use of a computer, or other device, in *real time* in such a way that the operator can control its activity.

interactive routine a programming *routine*

in which a series of operations is performed repeatedly, until a previously specified end-condition is reached.

interactive videotex see *videotex*.

interface used as a general term to describe the connecting link between two systems. Most frequently refers to the *hardware* and *software* required to couple together two processing elements in a computer system. Also used to describe the recommended standards for such interconnections. For example Interface-*CCITT* is an international standard for the interface between data processing terminals and communication equipment.

interference confusion, or loss of clarity, caused by unwanted signals, or *noise*, in a communication system.

interlacing in TV *scanning*, the technique of interposing two fields to reduce *flicker* (see the diagram under *flyback*).

Inter-library loan (ILL) the exchange of information in *hard copy* or *microform* between libraries.

interlock a computer device to prevent unauthorized access to, or change of, *data*, eg *log-in* procedure.

INTERMARC International Machine-Readable Catalogue (see *MARC*).

intermediate copy a copy of a document which serves as an intermediate stage in producing a final copy, eg a photographic white-on-black negative from which a black-on-white copy can be made.

internal memory *memory* to which the *central processor* of a computer has *direct access*.

International code another (usually American) name for Morse code.

International Economic Abstracts a Dutch *database* covering business and economics information. It is accessible via *Lockheed*.

International Electro-Technical Commission an international organization, based in Geneva, which recommends international standards for electrically operated equipment. It works with *ISO* in this specialized area.

International Frequency Registration Board (IFRB) one of the three main organizations within the *International Telecommunications Union* (ITU). IFRB is particularly involved in registering and standardizing the radio frequencies used internationally.

International Institute of Communications UK-based, but with international trustees, officers and Advisory Council, IIC analyses social, political, cultural and legal issues relating to communication (particularly electronic communication). It aims to assist governments and industries in the formulation of policies for handling new information technology.

International Nuclear Information System see *INIS*.

International Packet Switching Service IPSS is a public automatic *switched data* service providing access between UK *data terminals* and *computer* systems abroad (and vice versa). The service features *packet* assembly and transmission of *data*, and leads to *compatibility* between otherwise incompatible equipment (see *packet switching*).

International Special Committee on Radio Interference committee set up by the *IEC* to establish standards for telecommunications equipment, with particular reference to the control of radio interference.

International Standards Organisation a body which attempts to establish international standards and to help coordinate national standards.

International System see *SI*.

International Telecommunications Union (ITU) the ITU is the telecommunications agency of the United Nations. It has three main components – *CCIR*, *CCITT* and *International Frequency Registration Board.*

interpreter a computer program which controls the execution of another program which has not been previously *compiled* or *assembled.*

intersatellite link the transmission of messages between *communications satellites* (rather than from satellite to *Earth station*).

Intrafax *closed circuit facsimile transmission* system leased to US Government Agencies, military and industrial corporations by *Western Union.*

invariant field see *field.*

inverted file a form of *file* organization frequently used for *databases* in *information retrieval systems.* When *items* are added to a database, their attributes are identified by means of *index terms.* Such index terms are then brought together to create an inverted file, so called because it lists all items possessing that attribute. Thus, for example, in a *bibliographic database* an inverted file might be set up for all items (here titles of articles) containing the word 'oxygen'; or on a police records database, an inverted file might be created for all items (here persons) who had ever been charged with theft.
Inverted files occupy more storage space than is required for *serial files* holding an equivalent amount of data. But the creation of inverted files obviates the need for *serial searching,* and so greatly reduces *access time* (particularly important for *on-line searching*).

I/O an abbreviation of *input/output.*

IOB Inter-Organization Board for Information Systems and Related Activities, UN.

I/O buffer a temporary *storage* area for computer *input/output.* (See also *buffer.*)

IOC (or I/OC) *input/output controller.*

IOCS *input/output control system.*

I/O equipment all equipment which *reads input* or *writes output.*

IP *information provider.*

IPA International Pharmaceutical Abstracts. A US *database* accessible via *Lockheed.*

IPC industrial process control (see *numerical control*).

IPG Information Policy Group of the Organisation for Economic Cooperation and Development (OECD).

IPL 1. *international program loading.* 2. information processing language (see *language*).

IPSS *International Packet Switching Service.*

IR information retrieval (see *information retrieval system* and *information retrieval techniques*).

IRE Institute of Radio Engineers, US

IRL information retrieval language (see *information retrieval system, on-line searching* and *search language*).

IRMS Information Retrieval and Management System provided by *IBM,* US.

Iros a US *teleordering* system.

IRRD International Road Research Documentation. A *database* produced by the Organization for Economic Cooperation and Development (OECD) in Paris. It covers roads, traffic, vehicles, traffic safety and related topics. It is accessible via *SCANNET.*

IRT Institute of Reprographic Technology, UK.

ISAR information storage and retrieval (see *information retrieval systems*).

ISBD International Standard Bibliographic Description. An internationally used convention for the description of documents, established by the *ALA.*

ISBD (CM) *ISBD* for cartographic materials.

ISBD (G) general *ISBD*.

ISBD (M) *ISBD* for monographs.

ISBD (NBM) *ISBD* for non-book materials.

ISBD (S) *ISBD* for *serials*.

ISBN International Standard Book Number. Each book published is allocated its own unique number (ISBN), consisting of ten digits. These are made up of: a group identifier (linguistic, geographical, national or other relevant group); a publisher identifier; a title identifier; and a *check digit*, used to help identify errors in transcribing the number. The ISBN is used to aid the location of a book within a library, or *information retrieval system*, and it can also be used in *teleordering*. (See also *ISSN*.)

ISDN *integrated services digital network*.

ISDS *International Serials Data System* (of *UNISIST*).

ISI Institute for Scientific Information: compilers of the *Science Citation Index* in the US.

ISL *intersatellite link*.

ISMEC Information Service in Mechanical Engineering. A US-based *database* covering all aspects of mechanical engineering. It is accessible via *ESA/IRS*, *Lockheed* and *SDC*.

ISO *International Standards Organization*.

ISR 1. information storage and retrieval (see *information retrieval system*). 2. Index to

Scientific Reviews: *current awareness service* based on the *Science Citation Index*.

ISSN International Standard Serial Number. Each *serial* title published is allocated a unique number (ISSN) consisting of eight digits (in two groups of four). The first seven digits constitute an unambiguous title number for the serial: whilst the eighth is a *check digit* used to help identify errors in transcribing the number. The ISSN is used to aid communication between publishers and libraries, to aid *inter-library loan* and *document delivery systems*, and in the maintenance of *bibliographic databases*. (See also *ISBN*.)

ISSP Information System for Policy Planning. A US Government system which serves the Office of Management and Budget. (Incorporates *DIDS*.)

ISTIM interchange of scientific and technical information in *machine language*.

IT *information technology*.

italic a sloping *type fount*.

item a unit of information relating to a single document, person, etc, contained within a *database*.

item size the number of *characters* in a unit of data.

ITM *information transfer module*.

ITU *International Telecommunications Union*.

IWP International Word Processing Organization: based in the US (see *word processing*).

J joule: *SI* unit of energy.

jacket has a specific meaning in *microform*, when it refers to transparent plastic envelopes used for the insertion of short strips of *microfilm*.

jack plug see *jack socket*.

jack socket a connecting socket used for terminating the wiring of a circuit. Access is gained by inserting a jack plug.

JAPATIC Japan Patent Information Center. Produces the *database*, and acts as *host* for the Japanese PATOLIS patent search service.

JCL *job control language*.

JICST Japan Information Center of Science and Technology. Japan's largest *database* producer and *host* (see *JOLS*).

job a specific piece of work to be *input* to a *computer*. Each job normally requires a number of *runs*.

job control language a language understood by a computer's *operating system*. It allows users to tell the computer how a *job* should be controlled within the system.

job-orientated language a computer *language* which is designed for the specific needs of a particular type of *job*.

job-orientated terminal a *terminal* designed for a particular type of *job*, eg checking and making flight reservations, conducting stock market transactions.

JOIS Japanese On-line Information System operated by *JICST*. The present service (JOIS II) enables users to retrieve *bibliographic* information *on-line* in Kana (Japanese symbols), Kanji (Chinese charac-

ters) or English from the *JICST* database. JOIS II also gives access to some major non-Japanese *databases*.

joystick a lever whose motions control the movement of a *cursor*, or it can be used to *write* on a *VDU*. The name derives from the analogous lever used to control the movements of an airplane.

J-tree see *junction grammar*.

junction grammar (JG) a model of language structure used for analysing sentences in *machine translation*. It allows each sentence to be broken down into its syntactic elements (subjects, verbal predicates, objects, etc). These are arranged as a structured array resembling a family tree (called a *J-tree*). The meaning of a sentence is then derived from the relationship between the meanings of individual elements in the *J-tree*. (See also *machine-aided translation* and *HAMT*.)

justification see *justify*.

justification range text is usually *justified* by introducing variable word spaces. The justification range defines the permitted minimum and maximum space which can be inserted between words in a line.

justification routine a computer *program* that enables a *phototypesetter* to produce *justified* material.

justify to adjust the positions of words on a page of text so that the margins are regular, with lines beginning (and/or ending) at the same distance from the edge of the page. Type can be aligned on the left, with a ragged right margin (flush left); or, less often, aligned on the right with a ragged left margin (flush right); or, as is customary in traditional printing, the type can be justified by aligning both margins.

J

k an abbreviation for kilo, denoting a thousand (10^3).
When referring to *storage capacity*, k is generally used to mean about a thousand. For example, a 64k *word* store actually contains 65,536 words.

Kansas City Audio Cassette Standard a standard convention for encoding data on *audio-cassette* tapes.

kb kilobyte.

kcs one thousand (k) characters per second. A unit of *data transmission* speed, eg 50 kcs = 50,000 *characters* per second.

kern in *hot metal* this is any part of the type that extends beyond the main body. In *phototypesetting*, an overhang between adjacent characters must be correctly overlapped. This process is called either 'kerning', or 'mortising'.

kerning see *kern*.

key 1. a marked button, or lever, which is depressed (or touched) to register a *character*. 2. a group of *characters* used in the identification of an *item*, and to facilitate access to it.

keyboard a device equipped with an ordered array of *keys* (1.), which are manually operated to encode *data* or instructions. Diagrams overleaf.

keyboard lockout a property of a *keyboard* such that the keyboard cannot be used to send a message over a *network* whilst the required circuit is engaged.

keyboard send/receive a *teletypewriter* transmitter and receiver which transmits from the *keyboard* only.

key letter in context see *KLIC*.

keypad a hand-held *keyboard*, used to provide electronic *input*, which has fewer keys than a normal *terminal* keyboard.

key phrase in context see *KPIC*.

key punch synonym for *cardpunch*.

key stroke the operation of a single key on a *keyboard*.

key-to-disk a data entry technique in which data is sent directly from a *keyboard* to a disk file.

keyword a substantive word in the title of a document (or other item within a *database*) which can be used to classify content. Such words provide access to the item when they are used as *search terms* (see *information retrieval systems*). For example, the keywords in the title, 'The Influence of Smoking on Lung Cancer', would be 'Smoking', 'Lung' and 'Cancer'.

key word and context see *KWAC*.

keyword in context (KWIC) a form of *automatic indexing*. As items are added to a *database*, *keywords* are extracted from their titles (or from *abstracts* or portions of text). Common words which are not indicative of content, eg and, of, the, etc, are eliminated by means of a *stop list*.
For each keyword a list can then be generated which shows the context within which it appears in each item.
An entry in a KWIC index normally consists of lines of text printed in such a way that the keyword appears in a central column (in alphabetical order), with context to left and right. A document reference number is entered for each line. (See, in contrast, *KWOC*.)

key word out of context see *KWOC*.

kilo a prefix signifying one thousand (10^3).

kilobaud a measure of data transmission speed: a thousand bits per second (see *baud*).

kilo stream a *digital data transmission* service offered by *British Telecom*.

KISS keep it simple sir. Computer jargon – refers to *programming* activities. ('Sir' is sometimes replaced by 'stupid'.)

KIT Key Issue Tracking. A US *database* which provides a constantly up-dated index of currently important topics.

K

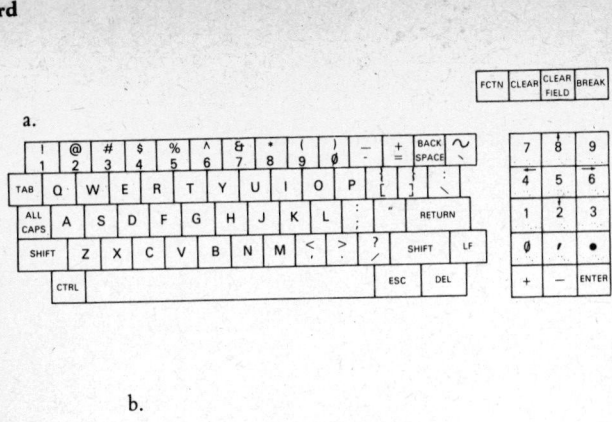

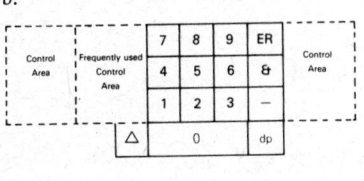

a.

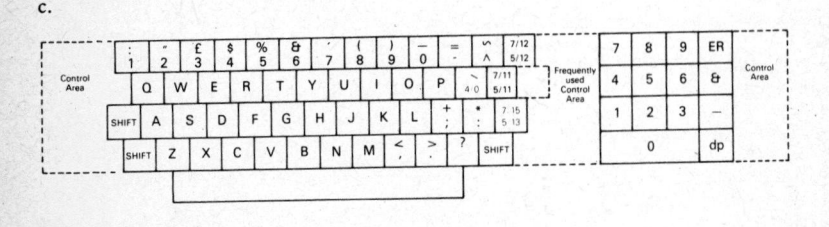

b.

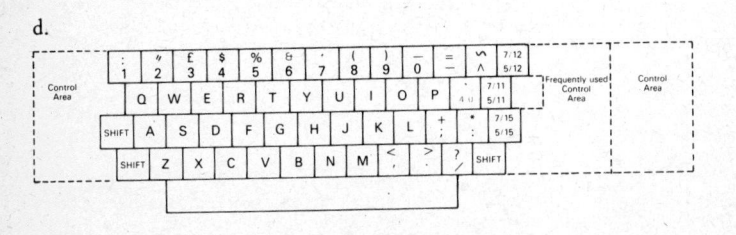

c.

d.

Four types of keyboard: a. A conventional, typewriter style keyboard for a word processor; b., c. and d., three keyboards from ECMA Standard 23, which shows keyboards generating the ECMA seven bit coded character set. (b) is used exclusively for numerical data; (c) is an alphanumerical keyboard for predominantly numeric data; and (d) is for predominantly alphabetic data.

KLIC key letter in context. An *indexing system* for producing *permuted* lists of terms. All terms are sorted by each letter in every term. The remainder of the term is also displayed. Similar to *KWIC* indexes, but based on letters instead of words.

klu(d)ge a slang term to describe an unsatisfactory combination of component parts in a system. Derogatory, possibly derived from German 'klug' (smart, witty).

KPIC key phrase in context. An *indexing system* similar to *KWIC* and *KLIC*, but using phrases instead of letters, or words, as the fundamental units.

KPO *key punch* operator.

KSR *keyboard send/receive*.

Kurzweil reading machine an aid for the blind, which converts printed matter into computer data using an *optical character recognition* (OCR) device. The data are then transformed to spoken form via *speech synthesis*.

KWAC key word and context. An *indexing system* similar to *KWIC*. Titles of documents are *permuted* to bring each significant word to the beginning, in alphabetical order. This *keyword* is followed by subsequent words in the title, and then by that part of the title which came before the significant word.

KWIC *keyword in context*.

KWOC key word out of context. An *indexing system* in which titles are printed in full under as many *keywords* as the indexer considers useful. (See, in contrast, keyword in context.)

label a *character*, or group of characters, used to identify an item of *data*, a *record* or a *file*.

Labordoc a *database* compiled by the International Labour Office in Geneva, covering labour, demography and related topics. It is accessible via *SDC*.

LAN *local area network*.

language a set of representations and rules by which information is communicated within, and between, computers, or between computers and their users. (See also *computer language*; *indexing language*; *microcomputer language*; *source language*; *target language*.)

language translator 1. a computer *program* used to translate other programs from one *language* to another, eg BASIC to FORTRAN. 2. a program which assists in the translation of spoken languages, eg from English to French. 3. a part of the computer which translates *input data* and *instructions* into *machine code*.

large scale integration see *LSI*.

LARP local and remote printing. An expression used in *word processing* to indicate that *print-out* can be obtained either locally (near the user's *terminal*), or at a more distant (ie remote) point.

LASER laser stands for light amplification by the stimulated emission of radiation. The laser was first developed in the 1960s.

It produces a narrow, high-energy light beam, which can be used for a wide range of communication activities. Examples are printing (see *electrophotographic printing*), *microform* production and optical scanning (see *optical character recognition* and *optical video disc*). The beam can also be used to carry signals along optical fibres.

lasercomp a device (manufactured by Monotype International) which uses *fount*, *stored* in *digital* form, and a horizontal scanning *raster* (acting on film or photosensitive paper) to record data.

laser electrophotographic printing see *electrophotographic printing*.

laser emulsion storage a *storage* medium based on the use of a *laser* to irradiate a photosensitive surface.

laser line follower an *input* device for graphic information. A *laser* beam follows and traces continuous lines, recording them in *machine-readable* form.
The laser line follower offers advantages over other forms of *scanner*, since it eliminates unwanted information, eg marks or dirt on the *hard copy*.

laser platemaker a device which can scan a page with a *laser* at one geographical location, transmit the resultant signals to a remote location, and produce a printing plate there using another laser. This is a form of *facsimile transmission*.

laser scanner the coherence of laser beams

L

mirror — glass tube — helium-neon gas mixture — resonator — mirror allowing partial transmission — coherent output beam — electrodes — high voltage power source

Layout of a typical (helium-neon gas) laser. By producing a coherent beam of light, very good definition is made possible in communications systems.

enables them to focus on a very small area. Laser scanners can therefore be used for both recording and reading data. This leads to their use for storage devices (see, eg, *laser emulsion storage*), and for the sensing of characters as in *optical character recognition*.

latency a *waiting time* associated with the delay in *accessing data* from a *storage* device.

lateral reversal an image which has been reversed left-to-right.

launch vehicle the rocket used to place a satellite in orbit.

layout character synonymous with *format*.

LC Library of Congress, US.

LCCC Library of Congress Computer Catalog.

LCCMARC Library of Congress Current *MARC* file. A *database* giving a catalogue of monographs from 1977. Includes *CIP* data.

LCD *liquid crystal display*.

LCMARC Library of Congress *MARC* files starting in 1968.

LCR longitudinal redundancy check: a procedure for checking errors in a computer *program*.

LDRI low data rate input.

LDX long distance xerox. A form of communication combining *facsimile transmission* with *xerographic* copying.

leader 1. the first *record* in a *file*, or the first *field* in a record, which is used to identify that particular file or record. 2. the blank piece of paper which precedes the data recorded on a *paper tape*.

Leadermart derives, in part, from Lehigh Automatic Device for Efficient Retrieval. A system for automatic *natural language searching*. Documents are reduced to noun phrases for searching purposes. English

language requests are entered, and those noun phrases which best fit the request are displayed for examination. Documents can also be ranked in order, depending on the extent to which they match the original search statement. (See also *BROWSER* and *SMART*.)

leaders dots (or dashes) used by the printer to fill in a line so as to lead the reader's eye across the page to data at the end of the line.

leading (pronounced 'ledding') a term (derived from traditional *typesetting*) for the insertion of space between lines of text. The amount of space is usually measured in *points*. In *phototypesetting*, the term is used for the advance of the film after each line is set.

learning curve the growth of individual productivity with experience. For example, operators using *word processors* for the first time will usually become increasingly efficient as their experience of the system grows. The name derives from the practice of plotting productivity gains graphically.

leased line a telecommunications link (most often a telephone line) reserved for the sole use of the leasing customer.

LED display *light emitting diode* display.

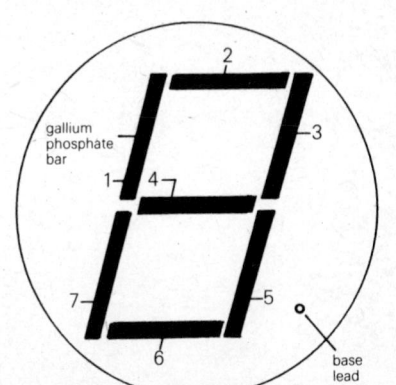

Example of an LED display element. To display, for instance, the numeral 6, terminals 1, 4, 5, 6 and 7 must be energized.

left justify sometimes used as the equivalent of *flush left* (see *justify*).

legal retrieval the use of an *information retrieval system* to obtain information on legal matters.

letterpress the main printing process for the last 500 years, but now in decline. Printing is by direct contact between an inked, raised image and the paper. There are three main types of press: plates, flatbed cylinder (where the paper is wound on a cylinder), and rotary (where the printing plate, itself, is cylindrical).

letter-quality printer any *printer* that can produce print of the same quality as a typewriter.

letter shift (capital shift) 1. shifting from lower to upper case when entering text at a *terminal*. 2. in *Murray Code*, a shift from figures to letters.

LEXIS 1. Lexicography Information Service. This is a West German *pure MAT* system offering translation between English, French, German and Russian. It is based on an *automated dictionary*, normally interrogated in *batch* mode, with *COM* and printed *output* (see *machine-aided translation*). 2. a legal *database* which can be searched *on-line* (see *on-line searching*).

LIBCON/E Library of Congress/English. A *database* covering English-language monograph records derived from the Catalog(ue) of the US Library of Congress. It is accessible via *SDC*.

librarian 1. person who controls a *library*. 2. a *program* used in maintaining a *library* (sometimes more explicitly called a *librarian program*).

librarian program see *librarian* (2.).

library as used in computer terminology, this refers to a collection of tested *programs*, *routines* and *subroutines*. Computer centres often maintain such libraries for internal use and external loan.

library routine a tested *routine* that is

kept in a computer *library*.

LIBRIS Library Information System. A Swedish *on-line information retrieval system*.

Libris a US *teleordering* system.

LIFO last in, first out. Computer jargon referring to a sequence followed in *data processing*.

ligature two or more characters which are joined together for printing purposes, eg fl.

light emitting diode a *diode* (electrical *switching* component) which emits light when excited by an electrical current.

light gun synonymous with *light pen*.

light pen an electronic *stylus*, containing a light sensor, which can be used to specify a position on a *cathode ray tube display*. Used for communication between a user and a computer, eg in *computer graphics*, page layout.

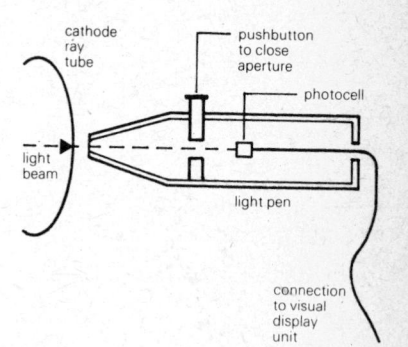

Schematic diagram of a light pen.

LILO last in, last out. Computer jargon referring to a sequence followed in *data processing*.

limited distance adapter a *modem* designed to operate over short distances (up to, say, 30 miles).

limiter a device used to reduce the power of an electrical signal when it exceeds a specified value.

Lindop Committee a committee established by the UK Government in 1978 to make recommendations for the safeguarding of information on computers. Its central recommendation was that there should be an independent *Data Protection Authority* to regulate the way computer *data* were handled, and to ensure that the privacy of the individual was protected.

line 1. any long narrow mark. 2. a *channel* through which signals can be transmitted. 3. a horizontal row of *characters* on a page, or *display screen*. 4. one scan across a *CRT* (especially in television).

linear predictive coding (LPC) a technique for analysing speech and converting it into *digital code*.

linear program(me) a form of programmed instruction which follows a predetermined sequence of (ever more complex) steps (see *computer-aided instruction*). Should not be confused with *linear programming*.

linear programming a mathematical technique for breaking problems down into a form amenable to *computer* solution.

line-at-a-time printer synonymous with *line printer*.

line concentration a means of matching a larger number of *input* channels with a smaller number of *output* channels – the latter usually working at a higher speed.

line drawing see *line illustration*.

line drawing display a *cathode ray tube display* on which lines can be drawn. The lines are either *input* directly using a *graphics tablet* or a *light pen*, or indirectly by defining the end points of the lines via a *keyboard*. Once entered, the lines can be extended, rotated or otherwise altered. Line drawing displays therefore provide a valuable tool in *computer-aided design*. (See also *computer graphics*.)

line driver see *bus driver*.

line-feed code a *code* which instructs a

printer to move the *platen* up one *line* (3.).

line finder a device attached to a *platen* which moves it automatically to print on a specified *line* (3.) of a printed form.

line flyback see *flyback*.

line (display) generator a device used in conjunction with a *cathode ray tube* to generate dotted, dashed or continuous lines.

line illustration an illustration made from drawn lines only.

line misregistration deviation of a *line* (3.) of *characters* from an imaginary horizontal base line.

line number 1. *photocomposed galley proofs* often have their lines numbered (as does computer *print-out* of text) to facilitate locating the position of corrections. 2. in some *high level programming languages*, eg BASIC, line numbers are used to order and locate each instruction.

line printer a device for printing computer output. The printer contains sets of characters on continuous belts (called print chains) which are driven by the computer. An entire line of characters is printed as paper is fed continuously past the printing head.

line speed the rate at which data can be transmitted over a communications channel. It is usually expressed as *bits per second*, or in *bauds*.

line status the status of a communication *line* (2.), eg whether it is ready to receive or transmit.

line switching *switching* where a *circuit* is set up between incoming and outgoing *lines* (2.). (Contrast with *message switching*.) The term is synonymous with circuit switching.

line termination unit alternative name for a *data set adapter*.

LIPL Linear Information Programming Language: a *high level programming language*.

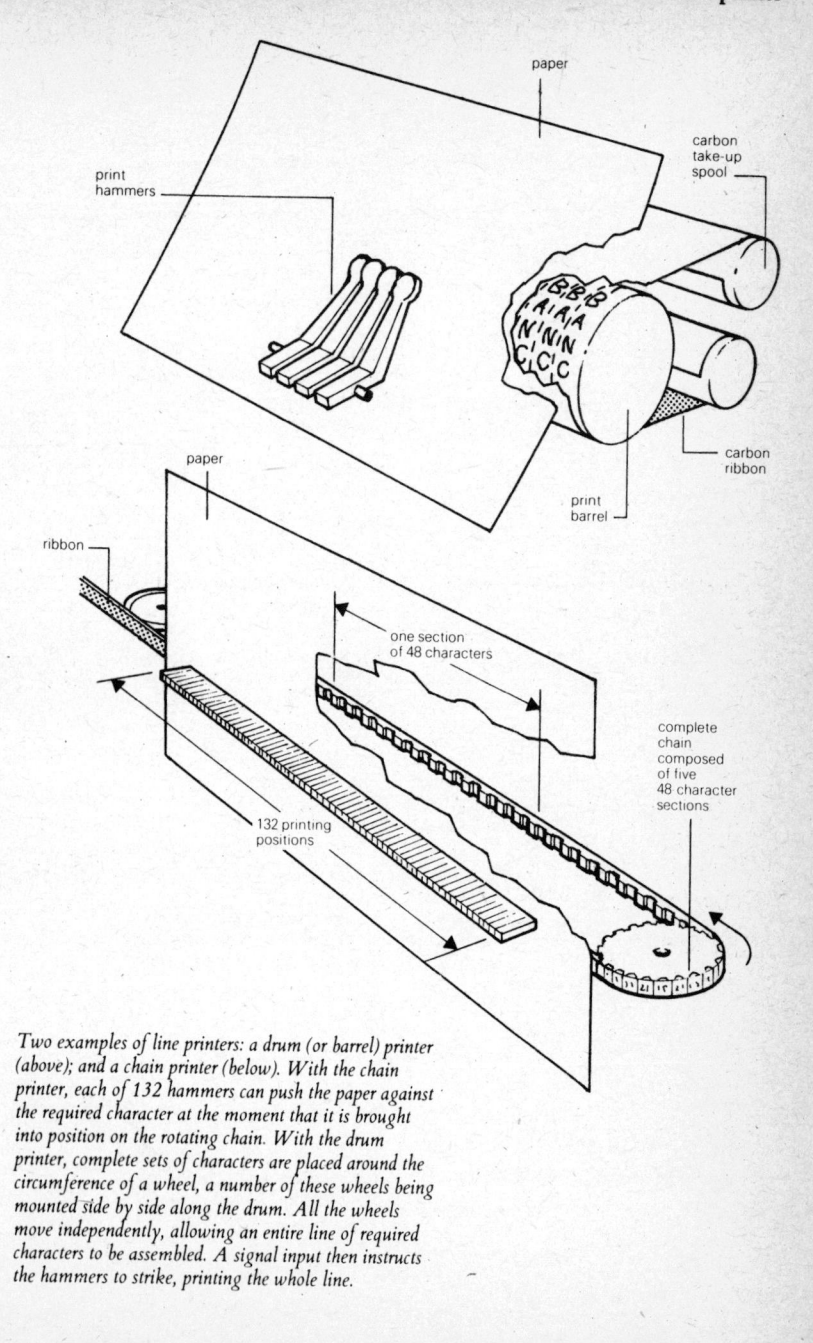

Two examples of line printers: a drum (or barrel) printer (above); and a chain printer (below). With the chain printer, each of 132 hammers can push the paper against the required character at the moment that it is brought into position on the rotating chain. With the drum printer, complete sets of characters are placed around the circumference of a wheel, a number of these wheels being mounted side by side along the drum. All the wheels move independently, allowing an entire line of required characters to be assembled. A signal input then instructs the hammers to strike, printing the whole line.

109

liquid crystal display liquid crystals do not generate light (unlike *LEDs*), but can be switched from an opaque to a transparent state. This can provide a data *display* if the liquid crystals are interposed between a light source and the observer.

LIS 1. Library and Information Science (see *information science*). 2. Lockheed Information Systems (see *Lockheed*).

LISA Library and Information Science Abstracts. A UK-based *database* accessible via *SDC*.

LISP List Processing: a high level language developed for *list processing*, and used for text manipulation.

LISR Line Information Storage and Retrieval. An information system used by *NASA*, US.

list 1. a series of *records* in a *file*. 2. the act of printing such a series (without performing any additional *processing*).

list processing *processing data* which are arranged in the form of *lists*.

literal 1. a symbol which defines itself, rather than standing for some other entity. For example, an integer is a literal, whereas a term in an algebraic expression is not. 2. in printing, it can mean a single *character* error.

literature search a search through a mass of documents to find those which can satisfy a user's requirements. The material sought usually relates to a particular specified topic, and the search is now increasingly frequently carried out on-line (see *on-line information retrieval*).

lithography a method of printing from a flat surface of stone or metal. The image areas on the plate are coated with a water-repellant ink, whilst the non-image areas are protected from the ink by a film of water.
Modern presses typically use a rubber-covered cylinder to transfer the plate image to paper. This process is therefore called *offset litho*.

liveware personnel involved in the running of *computers* and computer systems.

load to enter information, or a *program*, into a *computer*.

loader a program which controls the operation of peripheral units while other programs are being read into the computer's memory.

loading adding *inductance* to a *transmission line* to reduce *amplitude distortion*.

load on call loading data, or *programs*, into *storage* as, and when, they are required.

lobe the angular region over which an antenna (or *aerial*) experiences strong reception.

LOCAL load on call. This usage should not be confused with the normal use of the word 'local', as in '*local area network*' (see *load*).

local area network a system which links together *computers*, *electronic mail*, *word processors* and other electronic office equipment (see *work station*) to form an inter-office, or inter-site *network*. Such networks usually also give access to 'external' networks, eg public *telephone* and *data transmission networks*, *viewdata*, *information retrieval systems*, etc).

local central office see *local exchange*.

local exchange the *exchange* where individual subscribers' *lines* terminate.

local line the *channel* connecting a subscriber's equipment to the *local exchange*.

local loop see *local line*.

local network see *local area network*.

local origination in the context of *cable television*, this refers to television program(me)s which are produced within the local community.

location a *storage* position which can hold one *computer word*.

Location is designated by a specific *address*.

Lockheed the world's largest *database host*. It offers *on-line* access to over a hundred *databases* covering a broad spectrum of subjects.

logic when used in such terms as 'logic system' or 'logic chip', this refers to a system whose components can only take up one of two states. It is often taken to be synonymous with *binary* and *digital*.

logic unit see *arithmetic and logic unit*.

log in synonym for *log on*.

logo see *logotype*.

log on/off to initiate, or terminate *on-line* interaction with a *computer*.

logotype a trademark or similar small diagram.

loop a sequence of instructions within a *program* which are performed repeatedly until some predetermined condition is met. At this point, the computer exits from the loop, and proceeds with the next instruction in the original program. For example, the loop may contain instructions to repeat a series of numerical additions until a certain total is reached. At that point, the computer is instructed to leave the loop, print out the result and stop.

low activity data processing carrying out relatively few transactions on a large *database*.

lowercase (characters/letters) small letters of a *fount*, eg 'a' as opposed to 'A'.

low speed storage *storage* for which access is so slow that it limits the rate at which data can be processed. This implies that the access speed is slower than the *central processor's* calculating speed and/or the speed of *peripheral units*.

LPC *linear predictive coding*.

LPM *lines per minute*.

L-SAT an *ESA communications satellite* project for *direct transmission satellites* to provide television program(me)s in Western Europe and to establish the value of *digital* voice links. (See also *TV-Sat/TDF*.)

LSI large scale integration. An LSI circuit is one whose complexity is such that it would require more than ten thousand *transistors* to duplicate it. However, LSI is often used to refer to the production of such a circuit on a single silicon *chip*. (See also *ULSI* and *VLSI*.)

LSI memories *LSI chips* used as storage devices. They may be of a variety of different types: RAM, ROM, PROM or EAROM.

luminance a measure of brightness: especially applied to television signals.

m an abbreviation of milli: a prefix meaning one thousandth (10^{-3}).

M an abbreviation of *mega-* (signifying one million).

MAC machine-aided cognition (see *artificial intelligence*).

machine in information technology, 'machine' is often used as a synonym for *computer* (as in, eg *machine-code, machine language, machine-readable*).

machine-aided cognition see *artificial intelligence*.

machine-aided translation (MAT or pure MAT) the problems facing complete language translation by *computer* (see *machine translation*) are so formidable that no fully-automated machine translation systems have, as yet, been developed. Although any use of computers to assist translation might be described as machine-aided translation, this term is usually reserved for a specific type of system. In this, the computer aids a human translator by providing rapid translations of particular words and/or terms. Such aid is sometimes referred to as pure MAT, as contrasted with *pure MT* and *HAMT* systems.

The relative success of pure MAT, as compared with other fully, or partially, automated translation systems, can be traced to two factors. Firstly, it deals with the aspects of translation which give human translators most difficulty (unaided, they spend over half their working time consulting reference volumes). Secondly, it deals with those aspects of translation which are most easily automated.

The simplest MAT systems are basically computer-based multilingual lexicons. Translators can use these to determine problematic words and terms, before translating a source text (see, for example, *TEAM* and *LEXIS*). Other systems offer additional facilities. In particular, some allow *text editing* to take place on a *display* screen with a provision for words to be 'looked-up' *on-line* while the text is being processed. Multi-screen, or *split screen*, display facilities provide an additional aid to the translator in such systems as *TARGET*.

Further variations appear in the way that the lexicons are constructed. Some systems, eg TEAM and LEXIS, use *automated dictionaries* containing separate listings of roots and affixes. This avoids the problem of listing all variants of a given root, with consequent unwieldy growth of the *database*. However, such systems require a high level of expertise on the part of the translator, and can lead to problems of incorrect identification of roots. Other systems, eg *SMART*, therefore use *automated glossaries*, which only contain whole words.

Even with these lexicons, a translator may find that some terms cannot be immediately matched with the database. An additional means of access to the lexicon, based on approximate meaning, then becomes necessary. Such access can be provided by an *automated thesaurus*.

Automated thesauri do not solve the problems of words which have multiple meanings (see the example discussed under *machine translation*). Many systems, eg TARGET, TEAM and TERMIUM, therefore include a subject classification for entries, so that those meanings of terms which do not relate to the subject matter of the source text can be excluded.

MAT systems also have difficulty with terminology in the form of 'phrases', the meanings of which are not clear from their constituent terms. Consider, for example, the phrase 'random-access back projection in computer-assisted instruction'. '*Random-access*' and '*computer-assisted instruction*' may be included in the lexicon, but a further entry for the complete expression is likely to be needed for satisfactory translation. Such entries are included within some systems, eg TERMIUM, in the form of a *terminology bank*. Flexibility of access is made possible by means of a *permutation index*, which allows an entry to be identified by means of any of its constituent words. Diagram overleaf.

machine code the *code* used in *machine language*.

machine cognition see *artificial intelligence*.

machine cycle the time it takes for a sequence of computer events to be repeated.

machine error

Also refers to the sequence itself. This consists of two basic steps: a. the instruction phase (I – cycle), when *instructions* are brought from *memory*; b. the execution phase (E – cycle), when the instruction is carried out by the computer.

machine error an error caused by a malfunction of the *computer's hardware*.

machine hearing see *artificial intelligence*.

machine independent language a programming *language* which can be understood by a wide range of *computers*, eg *high level languages* such as *COBOL* and *FORTRAN*.

machine language the *language* used by a *computer* for communicating internally to its related parts: the language in which the computer performs its arithmetical and editing functions.

machine-orientated language a programming *language* in which each *instruction* has either one, or a limited number, of *machine code* equivalents. *Machine-orientated languages* therefore require little reprocessing before being 'understood' by a computer. However, they are less easy for users to learn and apply than *high level programming languages*.

machine readable capable of being read by a computer *input* device.

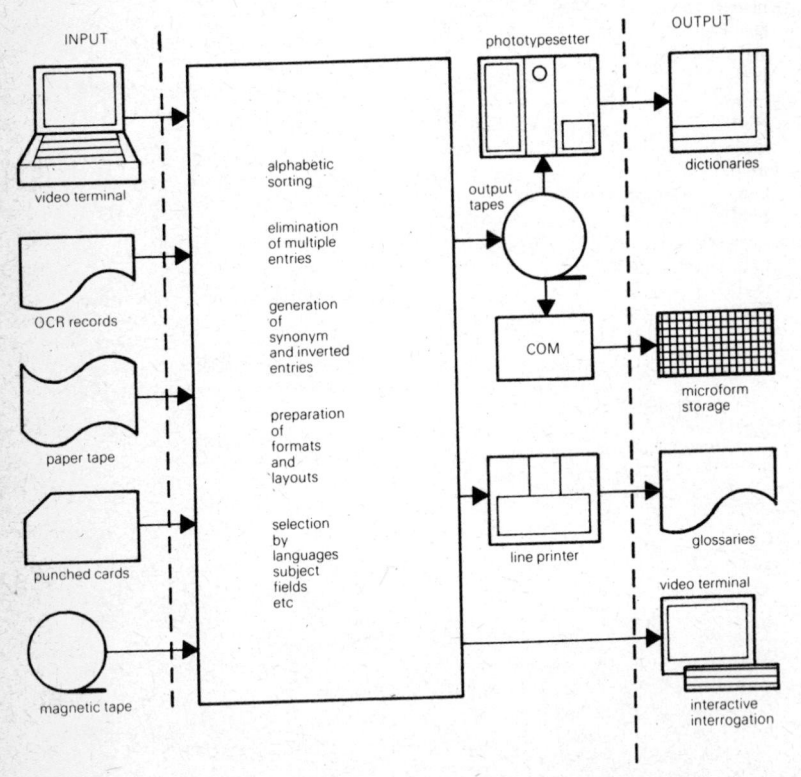

Typical example of a machine-aided translation system: the German TEAM system, which contains within the central computer an automated dictionary *and a* terminology bank.

machine translation techniques for automated translation of text from one spoken language to another have been sought since the Second World War. Computer technology (with its massive *storage* and *data processing* capabilities) has generally been seen as providing the basis for fully automated translation (ie translation without human intervention). However, a central problem has remained – how to relate the variable meanings of words to their context.

Take, for example, the word 'head'. It can be used as a noun, verb or adjective; it also has different meanings when used in the context of anatomy, numismatology, fluid dynamics, brewing, physical geography, football, analysis of social status, etc. The computer therefore has to recognize grammatical forms (remembering that verbs conjugate and nouns decline in both *source* and *target* languages) and subject contexts. It then has to deal with a further, and yet more difficult problem – that of recognizing the syntax of a sentence. The difficulties can be illustrated by comparing two simple sentences: a. 'He eats with his knife'; b. 'He eats with his wife'. In English, the preposition 'with', and its apparent place in the sentence, is common to both sentences. However, the human reader discerns that the syntax is different. This clearly has to be recognized by the computer, too, for different syntax may well require different prepositions and constructions in the target language.

These problems are such that the ideal of fully-automated machine translation (*pure MT*) seems unrealistic; at least in the immediately forseeable future. But two types of partially automated translation are currently proving themselves operationally viable: a. *machine-aided translation (pure MAT)* systems. In these systems, translations continue to be made by human beings, but they are assisted by computer-based aids, eg *automated lexicons, automated thesauri* and *terminology banks*; b. *HAMT* systems. In these systems, computers carry out the translation with assistance from human beings, who resolve syntactic and semantic ambiguities.

Pure MT systems are still being developed (see, for example, *SYSTRAN*). At present, they only give satisfactory results when supplemented by pre-input and/or post-output text processing, under the control of a human translator (see *machine-aided translation* and *HAMT*).

macro code a *code* which permits single *words* to generate several *computer instructions*.

macroelement a set of *data* elements handled as a unit.

macro-library a *library* of *routines* held in a mass *storage*.

macro trace an aid for *debugging* computer *programs*.

MAD Michigan Algorithmic Decoder: a *programming language*.

mag an abbreviation of *magnetic*.

Magazine Index an American *database* giving coverage of widely read popular literature. Approximately 400 of the most popular magazines in America are indexed. The database is made available via *Lockheed*.

MAGB *Microfilm Association of Great Britain*.

magnetic bubble memory a type of *storage* in which information is encoded onto a thin film of magnetic silicate in the form of bubbles. The presence, or absence, of a bubble in a particular location can be used to denote a binary digit. Detection of bubbles is carried out by means of a special sensor, which emits an electronic pulse as each bubble passes its *read head* (see *storage devices*).

magnetic card a card with a magnetizable surface upon which data can be recorded. (Used especially with some typewriters.)

magnetic card file a form of *backing storage*. Batches of *magnetic cards* are held in a magazine. When a data *location* on a card is called up, the card is transported at speed past a *read/write head*, giving *direct access* to the data.

magnetic card reader a device which can be used to input information from *magnetic cards* or transfer information from magnetic cards to another type of storage device.

magnetic card storage see *magnetic card file*.

magnetic cell a basic *storage* element in magnetic recording.

magnetic character a *character* which has been imprinted on a document using magnetic ink (see *magnetic ink character recognition*).

magnetic delay line a *delay line* whose mode of action depends on the time that magnetic waves take to propagate.

magnetic disc a form of *backing storage*. It consists of circular plates with magnetizable surfaces which possess a number of recording tracks, divided into sectors. Each track and sector is *addressable*, which gives disc storage its *random-access* capability. Several *read/write heads* enter and access data on each surface; each head covering a particular area. This provides *direct access* to data with *access times* of the order of 20-100 milliseconds.

magnetic disc unit see *disc unit*.

magnetic document sorter-reader see *magnetic ink character sorter*.

magnetic drum a form of *backing storage* giving rapid *direct access* (*access times* of a few milliseconds) and large capacity (in excess of 200,000 *words* for each drum, and several drums can be attached to a *central processor* at one time).
The drum is a cylinder with a magnetizable surface containing a number of recording tracks. The drum is continuously rotated, at speeds up to 7,000 revolutions per second while an array of *read/write* heads enter and access data with transfer rates as high as ten million *bits* per second.

magnetic film an *internal storage* medium consisting of a very thin (a few millionths of an inch) film of magnetizable material deposited on a plate of non-magnetizable material (usually glass).

magnetic ink character recognition the automatic recognition of *characters* printed with a special ink. The ink contains magnetic particles which can be detected and traced by appropriate input devices. Various standards are used for different purposes.

magnetic ink character sorter a machine which reads *magnetic characters*, and then sorts the documents upon which they appear. (Such machines are extensively used by banks to sort cheques.)

magnetic ink scanner an optical scanner capable of reading characters printed in magnetic ink.

(Left) magnetic ink character recognition characters, which are printed in special ink, picked up by input devices according to their shape on the grids seen here.
(Below) typical figures as used on bank cheque.

magnetic memory any *memory* using a magnetized material as the *storage* medium, eg *magnetic discs*, *drums*, *tapes*.

magnetic stripe system a magnetic stripe is applied to an object, and relevant information (usually about the object) is added. The data on such stripes can be read and updated, as necessary, using a *terminal*, or *wand*.
Magnetic stripes are extensively used in bank credit and *debit card systems* to identify a customer. They are also used to label products for sale in shops, so that details of the product can be recorded at the point and time of sale (an input to *automated stock control*).

magnetic tape currently, the commonest form of *backing storage* (and the cheapest form of magnetic storage). It consists of reels (normally 10½ inch diameter) of plastic tape (the commonest being ½ inch by 2,400 feet) coated with a magnetic oxide. Data are entered and accessed by means of *read/write heads* past which the tape is wound from one reel to another. *Direct access* is thus not possible; and all entries and searches have to be carried out in sequence. This limits the possibilities for *file* construction (see *serial file*), and gives slower *access times* than for other magnetic storage devices (*drums*, *discs*, etc). Different methods of recording data on magnetic tape are possible (see *NRZI* and *phase encoding*).

MAI machine-aided index (see *automatic indexing*).

mailbox service a name for *electronic mail*, especially applied to systems based on interactive *videotex*.

mailbox system refers to any system in which computer messages are transferred from one user (or organization) to the *file* of another, to await collection. The use of this term extends beyond *electronic mail*, to include, for example, the relay of document requests (made on completion of an *on-line search*) from a *host's* computer to that of a document fulfilment agency (see *ADRS*, *DIALORDER* and *Electronic Maildrop*).

mainframe sometimes refers to a compu-

ter's *central processor*, but is more frequently used to refer to any large computer, distinguishing it from a *minicomputer* or *microcomputer*.

make-up the preparation of a *page* for reproduction.

MAMMAX machine made and machine-aided index (see *automatic indexing*).

Management Contents a US *database* containing business and management information. It is accessible via *BRS*, *Lockheed* and *SDC*.

management information systems systems providing information for decision making, usually intended for senior management. The information may be internal to an organization, eg inventory levels, absenteeism, statistics, or external, eg commodity prices.

manual entry entering *data* into a *computer store* using a *keyboard*.

map a list of the contents of a *storage* device.

MAP Microprocessor Application Project. A project launched by the UK Department of Industry in 1978 with the aim of encouraging UK manufacturing industries to use *microelectronics* and *information technology* in its products and processes.

MARC Machine Readable Cataloging. A US Library of Congress system, developed by them and by the British Library, for the creation of *machine-readable bibliographic* records.

MARC (LC) Machine Readable Catalog (Library of Congress). A *database* made available via *BLAISE* and *SDC*.

MARC (S) Machine Readable Cataloging for Serials (see *MARC*).

MARC (UK) Machine Readable Catalogue (British Library, UK). A *database* made available via *BLAISE*.

Marecs marine *communications satellites*

under development by *ESA*. (See also *Inmarsat*.)

margin-adjust mode a term used in some *word-processing* systems. It refers to the facility to scan forthcoming *text* as an aid to justifying margins.

mark reading the reading of marks on a document using a *photoelectric* device.

mark sensing see *mark reading*.

mask in *word processing*, a form displayed on the screen with blank areas for the operator to complete.

mask matching a method used to determine the nature of *characters* during *optical character recognition*.

MASS multiple access sequential selection. A method of computer *data storage* and retrieval.

massaging manipulation of *input* material to produce the desired *format*, eg in *word processing*.

mass data an amount of data that is larger than the amount storable in the *central processing unit* of a given computer at any one time.

mass storage it is also referred to as '*backing storage*', or '*bulk storage*'. A large capacity store, such as a *magnetic disc*, used to supplement the smaller capacity, immediate access store in a computer system. The computer requires more time to *access* mass storage.

master film synonymous with *master negative film*.

master negative film an original negative, or a duplicate negative, reserved for the special purpose of making copies of *microform* for distribution.

MAT see *machine-aided translation* and *pure MAT*.

math(s) processing the capability, incorporated in some *word processing software*, which allows mathematical computations to be done.

matrix the master from which *type* images are formed in *phototypesetting*.

matrix printer a printer in which each character is composed of a series of dots produced by a stylus (or more than one) which moves across the paper.

MATV master antenna television. An *antenna* capable of providing television program(me)s in a similar way to *CATV*, but on a much smaller scale.

MB megabyte (= a million *bytes*). Used as a measure of storage capacity, eg of a *video disc*.

MCRS Micrographics Catalog Retrieval

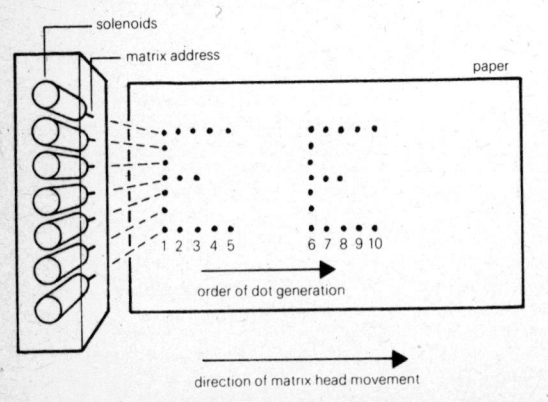

(Dot) matrix printer: the writing head moves from left to right across the paper, each character being built up by 'firing' the needles (of which there are seven) in sequence. Five such firing positions are required to generate a character, so one speaks of a 5 × 7 matrix.

solenoids

matrix address

paper

1 2 3 4 5   6 7 8 9 10

order of dot generation

direction of matrix head movement

System (of the Library of Congress, US).

MDO Marc Development Office (of the US Library of Congress – see *MARC*).

MDS multipoint distribution system. A *microwave* transmitter distributes television program(me)s to subscribers, who possess receiving *antenna* (usually mounted on the roof).

measure a printing term referring to the length of a line, eg in *picas*.

media-resident software *software* which is not an integral part of a computer system but is, instead, stored on some *medium*; normally a *magnetic disc*. (See, in contrast, *system-resident*.)

medium any material used to store information, eg *magnetic disc*. The word is often used in the plural – 'media'.

MEDLARS Medical Literature Analysis and Retrieval System. A *database* compiled by the US National Library of Medicine, which has offered a *batch retrospective search* service since the mid-1960s. An *on-line service* has been available since 1973 (see *Medline*).

Medline the *on-line* form of the *MEDLARS* database (the largest medical database, produced by the US National Library of Medicine). It is available in the UK via *BLAISE* (and, hence, via *Euronet*).

mega- prefix signifying one million (10^6).

Megadoc System a system developed by Philips, which uses *digital optical recording* (DOR) technology to record large numbers of documents on high capacity *DOR discs*. Can be utilized in an *electronic document delivery system* (such as *Adonis*). The name derives from '*mega* (= million) documents' (see *video disc*).

mega stream *digital data transmission* service offered by *British Telecom*.

memory a device into which *information* can be introduced and stored for extraction by a *computer* when required. *Direct access* is

required to the main memory of a computer. Such access must be rapid with *access times* independent both of the location of the information sought, and of the location of the last data elements accessed. *Magnetic core storage* and *thin film memory* satisfy these criteria, with *access times* of the order of *micro-*, or even *nano-*seconds. However, this storage is expensive (per data unit), and less costly *backing storage* is therefore used for large amounts of data. *Magnetic drums*, *card-files* and *discs* can be used to give direct access back-up, with *access times* of the order of *milliseconds*. *Magnetic tape* offers even cheaper memory capacity, but it does not offer *direct access*. Memory is often used as a synonym for storage.

memory capacity the amount of information which a *memory* element, or device, can store. It is also referred to as *storage capacity* (see *memory*).

memory map a guide to the location of various elements, eg *operating systems*, language *compiler*, graphics facilities, etc, within a computer's *memory*. Hexadecimal code is normally used to identify memory location (addresses).

menu a list of options, or facilities, from which a computer user can choose. A menu is usually displayed on the *VDU* of a *terminal*; a menu selection is made via the *keyboard*, or *keypad* or on some systems by pointing a light pen at the selected option.

menu selection making a choice from a *menu*.

Mercury UK *optical fibre* communication network currently in its initial stages.

MESH Medical Subject Headings. A *thesaurus* developed by the US National Library of Medicine, and used with *Medline*.

message a group of *words*, fixed or varied in length, which are moved about together as a unit.

message control flag a *flag* which indicates whether the information being transmitted is data or control information.

message header/heading the first *characters* of a message, indicating to whom the message is addressed, the time of transmission, etc.

message switching in telecommunications, the technique of receiving a message, storing it until the proper outgoing line is available, and then retransmitting. No direct connection between incoming and outgoing lines needs to be established. (See also *line switching*.)

messaging refers to a form of electronic communication in which a message is sent directly to its destination, eg *telex*. (See, in contrast, *store and forward*.)

METADEX Metal Abstracts/Alloys Index. These are *databases* compiled by the American Society for Metals and The Metals Society, UK. They are accessible via *ESA/IRS* and *Lockheed*.

metafont a *program* (devised by Knuth) which allows the operator to control the forms of *digitized type* used. The typeface style desired can be specified by an appropriate set of input parameters.

METAPLAN Methods of Extracting Text Automatically Programming Language is a *language* for text retrieval (see *text retrieval systems* and *information retrieval techniques*).

MF 1. *microfiche*. 2. *microfilm*.

MFLOP a million *floating point* operations per second. A measure of *computing power*, particularly in *supercomputers*.

MIC Medical Information Centre. A Swedish *host*.

MICR *magnetic ink character recognition*.

micro- 1. a prefix denoting one millionth (10^{-6}). 2. often used simply to denote smallness (as in *microfiche* or *microprocessor*). 3. often used as an abbreviation for microcomputer.

MICRO multiple indexing and console retrieval operations. An *on-line* system for retrieving, ranking and qualifying document references.

microcard an opaque card on which microcopies are reproduced photographically in rows and columns. It resembles a *microfiche*, but has its images printed positively. In consequence, the images cannot be directly reproduced.

microchip see *chip*.

microcircuit synonym for integrated circuit. (See *chip*.)

microcomputer a small (desk top) computer which uses a *microprocessor* as its processing element. Often used loosely to refer to the microprocessor itself.

micro copy a copy of an image, or document, so reduced in size (from its original) that it cannot be read by the unaided human eye (see *microform*, *microfiche*, *microfilm*).

microelectronics the use of integrated circuits in electronic devices.

microelectrostatic copying a system for updating *microform* using an *electrophotographic* technique.

microfiche a type of *microform* in which pages of text and graphics are photographically reduced and then mounted (usually as the negative) on a film. Each film usually has dimensions of 4 inches x 6 inches, and holds a matrix of 7 x 14 page frames. The fiche can be read a frame at a time using a *microfiche reader*.

microfiche book *hard-copy* text and plates are reduced to *microform* scale, and mounted on standard *microfiche*. The individual microfiches are then bound together in a hard, or soft, cover in such a way that they can be removed for reading. A lengthy book or report is thus reduced to a more convenient size for postage and storage.

microfiche management system a system for the storage and retrieval of information on *microfiche* (see *COM* and *information retrieval systems*).

microfiche reader an optical device for illuminating *microfiche* and providing an enlarged, readable image (usually projected onto a screen).

microfilm a type of *microform* in which photographically reduced pages are mounted, in sequence, on a roll of film.

Microfilm Association of Great Britain a body whose main aims are to promote the effective use of *microfilm* and raise the standard of microfilm production.

microfilm flow camera an automatic camera for taking microcopies of documents (ie recording documents on *microfilm*).

microfilm reader an optical device for illuminating a *microfilm* and presenting an enlarged, readable image on a screen.

microfilm reader-printer a *microfilm reader* which can produce (*blow back*) full-scale copies of the micro-images on demand.

microfolio a type of microform *jacket*. A number of strips of *microfilm* are placed adjacent to each other, and sheets of acetate film are attached on both sides. This produces a package which is about the same size as a conventional *microfiche*.

microform film in strips (*microfilm*) or sheets (*microfiche*) or opaque cards (*microcard*) onto which pages of photographically reduced documents are impressed. Some form of special magnifying device is required to read any type of microform.

microform in colour the vast majority of *microforms* are produced in monochrome, but colour *microfiche* and *microfilm* can be produced (usually for colour illustrations).

microform reader-printer a device for reading *microforms* which combines a viewing screen for visual display with a mechanism for producing enlarged print-on-paper copies of the material viewed.

micrographics the technique of reducing documents to *microform*.

microinstructions a single computer instruction representing a simple concept, such as 'add' or 'delete'

micromainframe a *microcomputer* having the *computing power* of a *mainframe*. Such a computer is predicted for the near future. For example, IBM announced in 1980 that they had reduced a main portion of the *central processing unit* of one of their mainframe computers to a single *chip*.

micron one millionth of a metre (10^{-6}m).

microopaque the opaque card used as a base for a *microcard*.

microphotographics see *micrographics*.

microprint a positive *microcopy*, photographically printed onto paper.

microprocessor a *central processor* in which all the elements are contained on a single *chip*. It is often used now as a synonym for *microcomputer*.

Microprocessor Application Project see *MAP*.

microprogram a program written in *microinstructions* to be implemented by *hardware*. It is not normally accessible to the computer user.

micropublishing the production and distribution of information via *microform*. The information may be original, but, in present practice, has often been previously published in conventional form.

microrecording a copying technique in which the copy is reduced so much in size that a special optical device is needed to read it. The copy may be on film (*microfilm*, *microfiche*), opaque card (*microcard*), or paper (*microprint*).

microrobotics the production of small handling devices (especially small robot arms) controlled by a *microcomputer* (see *robotics*).

microwave any radio wave with a frequency above 890 *megacycles per second*.

MIDAS Multimode International Data Acquisition Service. A *network* which allows *on-line searches* to be made from Australia of *databases* held by international *hosts*. The charging system is independent of distance. (See also *AUSINET*.)

midicomputers a computer of intermediate size and *computing power*, between a *mainframe* and a *minicomputer*.

milli- a prefix meaning one thousandth (10^{-3}).

Mind a pioneering experimental *HAMT* system which translates English into Korean.

minicomputer a computer of intermediate size and *computing power*, between a *mainframe* and a *microcomputer*.

MINICS Minimal Input Catalog(u)ing System. A document catalog(u)ing system.

minidisc a small (3-inch diameter) disc used for *storage* in some *word processing* systems and *microcomputers*. It can store roughly one page of A4 text. (Not to be confused with a *mini-floppy disc*.)

minification the scale of reduction of a *microform*.

mini-floppy disc a small (5¼ inch diameter) *floppy disc* able to store somewhat less than 100,000 *characters* (ie forty A4 pages of text). Double-density mini-floppies have greater storage capacity. These discs (also called *diskettes*) are used on *microcomputer* systems.

MIPS millions of instructions per second. A measure of *computing power*, particularly in general purpose *mainframe* computers.

MIS *management information systems*.

Missive French national *electronic mail* service. It will link together *telex*, *viewdata* terminals and automatic office *networks*.

Mistel a Finnish *viewdata* (interactive *videotex*) system.

MLA Modern Language Association Bibliography. A standard philological reference source, accessible via *Lockheed*.

MLS machine literature search(ing) (see *information retrieval systems*).

MMS *microfiche management system*.

Mnemonic code instructions written in concise, easily remembered symbolic or abbreviated form, eg SUB for subtract (see *assembly language*).

MOBL Macro Orientated Business Language: a *high level programming language*.

mode a particular method of operation, especially on a computer, eg *interactive mode*.

Modem an abbreviation of modulator-demodulator. A device for converting a *digital* signal (generated, for example, by a computer) into an *analog(ue)* signal by *modulation*. In this form, the signal can be transmitted along a standard telephone line. The received signal can be reconverted from analog(ue) to digital by the same device.

modern typeface printers use the term 'modern' to mean any typeface which shows a considerable contrast between the thick and thin strokes of letters.

modified NTSC *NTSC* television colo(u)r pictures played on *PAL* standard equipment.

modifier a quantity used to *modify* the *address* of an *operand*.

modify to alter the *address* of an *operand*.

modulation the addition of information to an *electromagnetic* signal (the '*carrier wave*'). Modulation can take the form of an adjustment to the carrier wave's amplitude (*amplitude modulation*), frequency (*frequency modulation*) or phase angle (*phase modulation*).

modulator a device that impresses audio or video signals onto a *carrier wave*.

module 1. any self-contained unit that can be added to (ie plugged into) an existing system. 2. a segment of core *storage* containing 20,000 *addressable* locations.

Molinya the name of a series of *communications satellites* launched by the Soviet Union. They have not generally been placed in *geostationary orbits*, but in elliptical orbits which provide maximum coverage of Soviet territory. In order to provide continuous communications coverage, it is therefore necessary to have a sequence of satellites following each other across the sky. In the 1970s, the Soviet Union began to launch Molinya satellites into geostationary orbit. The satellites in the Molinya series have been intended primarily for communication within the Soviet Union (see also *Statsionar*).

monitor 1. *hardware* or *software* used to monitor a computer system to detect deviations from some prescribed condition. 2. a synonym for an *operating system*. 3. a *cathode ray tube* used to monitor the quality of a television picture.

monochrome single colo(u)r, usually black and white but can be some other colo(u)r (especially with reference to television).

monospaced characters letters all of the same *set width* (as on an ordinary typewriter). This should be contrasted with *proportional spaced characters*.

mortising in *phototypesetting*, this means removing film with incorrect text and replacing it by film with correct text.

MOS metal oxide semiconductor field effect transistor. A very small, low power transistor which facilitates high packing densities in *integrated circuits* (see *N MOS* and *P MOS*).

mother board a large circuit board into which can be plugged a number of smaller boards, or circuit elements.

mouse a device which an operator can move over the surface of a *graphics tablet*. Its position is recorded by the computer, and can be used in moving text and illustrations about.

MPS *microprocessor* system.

MPU *microprocessor* unit.

MRDF *machine-readable data files*.

MTBF mean time between failures: a measure of equipment reliability.

MTBM mean time between maintenance. This provides a guide to *hardware* performance.

MTST magnetic tape selectric typewriter. A typewriter providing some editing facilities (see *word processor*).

multi-access system a system which allows a number of users to *access* a *central processor* in *conversational* mode at virtually the same time.

multidrop line a *line* (or *circuit*) which connects several *stations*.

multifunction system a computing system which can carry out a variety of tasks, eg a *word processor* with communication and mathematical capabilities.

multi-layer microfiche the normal 4 inch x 6 inch *microfiche* carries 98 'pages'. A long report or book can therefore require a large number of fiches.
Multi-layer microfiche offers a means for reducing this number. Two-layer microfiche can be made using polarization techniques, and four-layer microfiche using *holography*.

multiple access the ability of a system to receive messages from, and transmit them to, a number of separate locations.

multiplex to transmit two or more messages simultaneously via the same *channel*. This is achieved either by splitting the transmitted *frequency band* into narrower bands (frequency-division multiplexing), or by allotting the channel to several different inputs successively (time-division multiplexing).

multiplexor a device which uses and controls several communication channels simultaneously, both sending and receiving messages.

multipoint line see *multidrop line*.

multiprocessing the simultaneous execution of two, or more, computer *programs* on a *multiprocessor*.

multiprocessor a computer with multiple *arithmetic* or *logic units* that can be used simultaneously.

multiprogramming refers to the handling of more than one task by a single computer.

Murray Code (see *Baudot Code*).

NAK negative acknowledgement. A signal indicating to a sender that a previous message, or *data stream*, is unacceptable, and the receiver is ready for a repeat transmission. (See, in contrast, *ACK*.)

NAM *network access machine.*

NAND a NAND *gate* performs the logical converse of an *AND* function: not giving out a signal, if a signal is received along all the channels. A NAND gate with two inputs produces outputs as follows:

Input 1	Input 2	Output
1	1	0
0	1	1
1	0	1
0	0	0

nano- prefix denoting one thousand millionth (10^{-9}).

narrowband refers to a *bandwidth* of up to 300 *Hz*. (See in contrast *voiceband* and *wideband*.)

NASA the National Aeronautics and Space Administration in the US. One of its activities is to compile a large *database* covering all areas of aeronautics, space science and related technology. Part of the *database* (*IAA*) is publicly available via *ESA/IRS*. The remainder (*NASA STAR*) is only available to accredited users.

NASA STAR *NASA* Scientific and Technical Aerospace Reports. A *database* produced by NASA, and accessible *on-line* to accredited users.

National Aeronautics and Space Administration see *NASA*.

National Data Processing Service a division of *British Telecom* offering *data processing* and *transmission* services (and consultancy) to commercial customers. Its areas of specialization include *business information systems*, *Prestel*, *COM* and *information retrieval systems*.

National Library of Medicine a major centre for medical information in the US. Acts as a *host* and also compiles *databases*.

National Microfilm Association an association in the US whose main aims are to promote the use of *microfilm* and to improve standards of production. (See also *National Micrographics Association*.)

National Micrographics Association a trade association in the US representing producers of *microforms*. It has several committees dealing with standards for such areas as *microfiche* of documents, terminology, *information storage and retrieval*, newspapers on microfiche, reduction ratios, and *COM* format and coding. (See also *National Microfilm Association*.)

National Register of Microform Masters a register of *masters* located at the US Library of Congress (see *microform*, *COM* and *master negative film*).

National Reprographic Centre for documentation a UK organization devoted to the study of *reprographics*. It publishes information, and provides both equipment evaluation reports and an inquiry service.

NATO Codification System a system established by member countries of the North Atlantic Treaty Organization to provide a unique identification code for any item of military equipment (see *classification and coding systems*).

natural language ordinary spoken, or written, language. It is to be contrasted with a *programming* or *machine language*.

natural language system an *information retrieval system* in which the *index terms* are words actually used in the document. Indexing by natural language is generally cheaper than by ascribing index terms from an *authority file* or *thesaurus*.

NBM non-book materials.

NBS National Bureau of Standards, US.

NBS-SIS *National Bureau of Standards – Standard Information Services.*

NDPS *National Data Processing Service.*

N

NDRO non-destructive read out (see *non-destructive reading*).

negative an image (usually photographic) with *tones* which reverse those of the original.

negative acknowledgement see *NAK*.

NEPHIS Nested Phrase Indexing System. An automated *permuted* subject indexing system (see *automatic indexing* and *indexing systems*).

nest 1. a *subroutine* embedded within a *program*. 2. a *block* of *data* embedded within a larger body of data.

nested phrase indexing see *NEPHIS*.

network in general, this term refers to a set of components connected by channels. In the context of information technology, it usually refers to a system of physically dispersed computers interconnected by telecommunications channels, eg *Euronet*.

network access machine (NAM) a *computer programmed* to help a user interact with a computer *network*, eg a network connecting a series of *host* computers. The NAM generates the specific procedures necessary to gain access, and then allows the information to be sought in a uniform way (ie it performs the necessary translation of commands). The user gains access via a *terminal* to the NAM.

network planning a management technique for scheduling and controlling large projects. It is more often called *critical path method (CPM)*, or *PERT*.

neutral transmission a method of transmitting *teletypewriter* signals which employs a two-state (on/off) signal. (See also *polar transmission*.)

new information technology the contents of this book. Distinguished from *information technology* only in the sense that the latter may include applications which are in no way related to *electronic* handling.

Newspapers On Microfilm the published

microfilm records of the US Library of Congress holdings covering both US and foreign newspapers.

New York Times Information Bank a *database* containing detailed *abstracts* from the New York Times and over sixty other major newspapers and magazines. *On-line* and *off-line* search facilities are offered. Full texts of New York Times articles are available to subscribers in *microfiche* form.

NEXIS a US *database* offering full-text of newspapers and agency news items. It can be searched *on-line*. Similar in concept to the *LEXIS* legal database.

nexus a point in a system at which interconnections occur, eg the meeting point of a number of *channels* in a *network*.

NFAIS the National Federation of Abstracting and Indexing Services, US.

nibble computer jargon for a block of 4 *bits* (ie half a byte).

NIM Newspapers In Microform (see *microform*).

NLM *National Library of Medicine*.

NMA 1. *National Microfilm Association*. 2. *National Micrographics Association*.

N MOS N-channel metal oxide semiconductor. A common form of *transistor* (where 'N' stands for negative). N MOS is more frequently employed than *P MOS*. Electronic circuits based on MOS are relatively slow, but they are compact and consume little power.

noise 1. refers, in telecommunications, to unwanted (usually random) electronic signals. 2. refers, in information storage and retrieval systems, to retrieved documents which do not deal with the required subject. (See also *signal-to-noise ratio*.)

noise killer an electrical device used to reduce *interference* from a *transmission*.

non-destructive reading *reading* in which the source *data* are not destroyed.

non-destructive read out process in which *data* are obtained ('read out') from a *file*. The data in the file meanwhile remain unaltered in the computer's *memory* or *backing storage*.

non-volatile memory computer *memory* which preserves data during a loss of power to the computer, or a shutdown of the system.

NOR a NOR gate performs the logical converse of an *OR* function between input states (so inverting the output).

Notepad an American *teleconferencing* system.

NRCd *National Reprographic Centre for documentation*, UK.

NRMM *National Register of Microform Masters*.

NRZI non-return to zero indicator. A method of recording binary *digits* on *magnetic tape*. A change in direction of magnetization corresponds to 0. (See also *phase encoding*.)

NTIA National Telecommunications and Information Agency. An agency of the US Department of Commerce.

NTIS National Technical Information Service. A US service which generates a multi-disciplinary *database* covering technical and scientific reports produced by US Government agencies and their contractors. It is accessible via *BRS*, *CAN/OLE*, *ESA/IRS*, *Lockheed* and *SDC* and supplies copies of reports in microform.

NTSC National Television System Committee. A US body responsible for the specification of the US colo(u)r television system. The system is also used in Japan and South America (see *PAL* and *SECAM*).

null character a *character* employed either to fill unused time in a *data transmission*, or to provide a space in a *storage device*.

numerical control the automatic control of machinery (especially of production systems) by means of numerical instructions.

oblique in *phototypesetting* refers to the possibility of adding an angle of slant to a *typeface*.

OCCS Office of Computer and Communication Systems of the US *National Library of Medicine*.

Oceanic Abstracts a *database* covering oceanography and related topics, eg fisheries and pollution. It is accessible via *ESA/IRS*, *Lockheed*, *QL* and *SDC*.

OCI Office of Computer Information. An office of the US Department of Commerce.

OCLC Ohio College Library Center. An *on-line computer network*, mainly in North America, for catalog(u)ing information. It has a variety of library uses, eg *inter-library loan*, and locating information. Over 3,500 remote computer *terminals* are linked into the system.

OCR *optical character recognition*.

OCR-A a special typeface used on documents intended to be read by *OCR*. Presents a rather clumsy appearance to the eye. (See also *OCR-B* and diagrams overleaf.)

OCR-B a special typeface used on documents intended to be read by *OCR*. Pleasanter to the eye than *OCR-A*.

octal a number system based on the base 8 (ie it consists of digits 1 to 7).

octal (code) a *code* which operates with a base 8. (Compare with *binary*.)

ODB output to display buffer. An auxiliary *storage* area in a computer. It holds data during their transmission from *output* to a display *device*. (See also *buffer*.)

odd-even check see *parity check*.

OEM original equipment manufacturer(s). A term used when procuring compatible equipment for an existing computer system.

office information system any electronic system which can help perform a variety of office functions, including *word processing*, *information retrieval* and *telecommunications*.

office of the future see *electronic office*.

office of tomorrow see *electronic office*.

off-line not having a direct interaction with the *central processor* of a computer.

offset litho(graphy) see *lithography*.

Ohm *SI* unit of electrical resistance.

OIS *office information system*.

old style a *typeface* design with less contrast than *modern style*.

OLRT on-line real time operation (see *on-line* and *real time*).

on-line refers, in general, to any use of equipment to interact directly with the *central processor* of a computer.

on-line editing the manipulation of information stored in a computer using access via a *terminal*.

on-line information retrieval systems see *information retrieval system* and *on-line searching*.

on-line journal see *electronic journal*.

on-line remote microfiche reader a central bank of *microfiches* is indexed and coded for retrieval. The index can be searched *on-line*, and a particular fiche requested. A user can select and view any desired frame from a remote terminal, using a *video monitor* and *keyboard*.

on-line retailer see *host*.

on-line retrieval see *on-line searching*.

on-line searching this describes the activity of using a computer-based *information retrieval system* when there is direct *on-line* access to the *database*(s) available on the computer (or *computer network*). In on-line searching, the computer responds to a series of queries from the user. The latter starts

O

GB £ NET +001.50

ISBN 0-600-20346-8

00150

9 780600 203469

Practical application of two different optical character recognition systems, for information on a paperback book. The universal product coded bar-code is used to represent both price and ISBN number, these sets of figures also being reproduced in OCR-readable characters above.

0123456789
ABCDEFGHIJKLM
NOPQRSTUVWXYZ
abcdefghijklm
nopqrstuvwxyz
*+-=/.,:;"'_
?!()<>[]%#&@^
¤£$¦¡\
ÄÅÆIJÑÖØÜ
åæijøß§¥
"´` ^ ~
 ,
{}m_

Examples of OCR-B character set

by selecting appropriate search terms (ie the indexing terms used for the database) from a dictionary (which can normally be called up on-line). The computer responds by listing items identified by these search terms and/or presenting more highly focussed search options. The search can then be extended until an acceptable number of the most suitable items has been located. Relevant items can either be printed out at the user's *terminal*, or, if the number of items is too large, *off-line* at the computer, whence they are dispatched to the user. *Bibliographic databases* deliver references and, sometimes, *abstracts*. The full text of documents is currently only provided by a few systems as print-out, but some allow a photocopy or *microfiche* to be ordered via the terminal.

The entry point to such a retrieval system is typically a *video terminal* or an *on-line typewriter*. With an on-line typewriter, the systems response is printed on the paper output; while, with a video terminal, messages are shown on the screen and a printer is used to make a *hard-copy* record.

Throughout the 1970s the number and size of databases available on-line has grown. Simultaneously, accessibility has been improved, and the costs of access reduced, by linking computers via *time-shared* telecommunications channels. Access to most databases is normally obtained through information retailers, or *hosts*, the largest of which are currently *Lockheed* and *SDC* (System Development Corporation of Santa Monica, California) in the US. There are many smaller hosts in Europe, several of which are linked via *Euronet*. (This was established to bring European systems and users closer together, and to protect the European on-line information market.)

ON-TAP On-line Training and Practice. A *database*, accessible via *Lockheed*, which contains extracts from other *databases*. It is designed to be used as an aid for the training of personnel in the techniques of *on-line searching* (no *off-line* printing facility is offered).

opaque screen a diffuse reflecting surface for viewing a projected image, eg of a *frame* of *microform*.

open system 1. a system which allows a variety of different computers and *terminals* to work freely together. 2. a system to which access is publicly available.

Information flow in an on-line searching system.

open system interworking the establishment of links between discrete computer systems and *networks* to create a freely interacting *open system (2)*.

open wire an electrical conductor supported above the ground surface, eg telephone wires on poles.

operand any one of the quantities entering into, or arising out of, an operation. In computing, it typically refers to the items, eg data, *addresses*, on which the computer is operating at the time.

operating system *software* contained in a computer which allows it to control the sequencing and processing of *programs*, and so respond correctly to user requests, eg to store a *file* of *data*, to compile and run a *program*.

operational amplifier a device which both amplifies and operates on an input *signal*. It is used in *analog(ue) computers*.

operator 1. a person who operates a *computer*. 2. a *character* which designates the operation to be performed (eg $+$, $\div$).

Opitz Classification System a German classification and coding system for engineering tools and components. Their shape and significant characteristics are described by a 5-digit number, while a second 4-digit number gives supplementary information on dimensions, materials, etc (see *classification and coding systems*).

OPM operations per minute.

OPR optical *pattern recognition*.

optical cable see *optical fibre (fiber)*.

optical character recognition a technique in which information recorded on *hard copy* is examined by an *optical scanning device*. This *optical reader* converts the scanned information into *digital* form, so that it can be handled subsequently by electronic means. The input *characters* that can be accepted by the scanner are currently limited (see *OCR-A* and *OCR-B*). But the range is expanding rapidly to include

handwriting, etc. Diagram on page 130.

optical digital disc see *video disc*.

optical fibre (fiber) a very thin flexible fibre of pure glass. It can carry as much as a thousand times the information possible with traditional copper wire. A large number of fibres can be packed together into flexible cable. There are two main light sources used for optical fibres; *lasers* and *light emitting diodes* (LEDs). Lasers have some advantages, especially over long distances. They have narrower spectral width (ie the light they produce varies less in *frequency*), and they have greater 'launch' power into the fibres. However, LEDs tend to last longer, and are more stable over some *bandwidths*. (See also *fibre optics*.)

optical reader a device which can *read data* from a card, or document, using optical techniques.

optical scanning device a device which scans text or *graphics* and generates *digital* representations for computer processing (see *optical character recognition*).

optical transmission use of the visible part of the electromagnetic *spectrum* for communication. Currently, two main methods predominate: a. non-coherent transmission. Typically, a *light-emitting diode* (LED) is used which emits light when a current is passed through it. This is observed at the reception point by a suitable photodetector. This method of transmission is normally used over short distances (a few hundred metres); b. coherent transmission. The transmitting device here is a *laser*, which can provide a larger *bandwidth* over longer distances.

optical type fount a *fount* which can be read both by machines and by the human eye.

optical video disc see *video disc*.

optoelectronics the use of electrical energy to generate optical energy, or vice versa (see *optical fibres* and *light emitting diodes*).

refractive
index profile

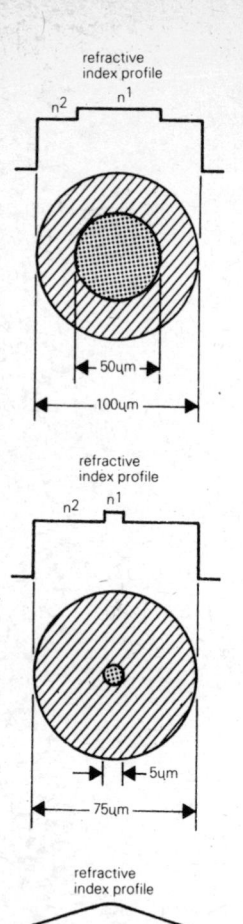

Rays are trapped within a glass core surrounded by glass of a lower refractive index. Multiple beams travel down the fibre, slightly dispersing the signal as they have differing path lengths.

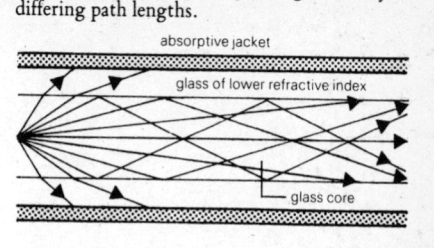

By using a smaller core of the same type as above, the light is effectively forced to travel purely as axial rays. Low dispersion results, allowing a high bit rate to be transmitted, but a laser is needed to inject sufficient light into the small diameter core.

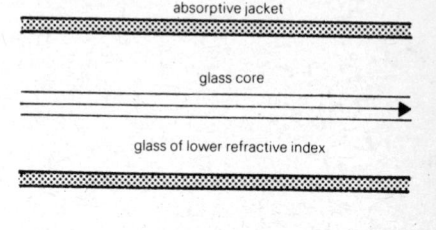

A refractive index made variable across the cross-section of the glass produces continuous re-focusing of the rays. This allows very high bit rate transmission.

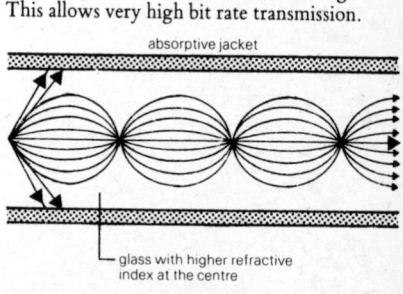

Three types of optical fibre transmission of light.

OR an OR *gate* is used in computer *logic* to combine *binary* signals in such a way that there is an output signal if any input channel carries a signal. For the case of two input signals, this leads to the following table.

ORACLE the *teletext* system of the British Independent Television Authority in the UK. The name is an acronym for optical reception of announcements by coded line electronics.

ORBIT On-line Real Time, Branch Information. A *computer language* developed by *IBM* and used by *SDC* for its *on-line databases* (see *SDC* and *on-line searching*).

Orbital Test Satellite an experimental *communications satellite* program begun by *ESA* in the latter part of the 1970s. (See also *European Communications Satellite*.)

ORION on-line retrieval of information over a network (see *information retrieval system* and *on-line searching*).

OSD *optical scanning device*.

OTP Office of Telecommunications Policy, US.

OTS *Orbital Test Satellite*.

output information transmitted by a computer, or its storage devices, to the outside world. It may, for example, be in the form of print-on-paper, *punched cards* or *paper tape*.

output bound a system whose speed of performance is restricted by the capabilities of the *output* system, eg a computer system might be output bound by a slow *printer*.

output device any device capable of receiving information from a *central processor*. It may be some form of *backing storage*, or a *peripheral unit* which 'translates' information into another medium, eg a *line printer* or *VDU*.

output limited see *output bound*.

overstrike the substitution of one *character* for another on a *visual display unit*. For example, in *word processing*, the *cursor* may be positioned below the character to be changed, and the desired character substituted via the *keyboard*.

overwriting the input of information into a computer so that it destroys information previously held in the same *location*.

PA paper advance. The movement of paper through a *printer*.

PABX private automatic branch exchange (see *private branch exchange*).

PAC personal analog(ue) computer (see *analog(ue) computer* and *microcomputer*).

PACC product administration and contract control. A concept applied in business *data management systems*.

package a generalized *program*, or set of programs (*software package*), written to cover the requirements of a number of users.

packet assembler/disassembler see *PAD*.

packet radio a technique envisaged for communicating with computers using small radio units. These may be either hand-held, or attached to computer *terminals*, and are able to transmit and receive short bursts of data.
The concept was devised by researchers working on *packet switching networks*, but is not limited in its application to such networks.

packet switched (or switching) a method of routing data, or a message, from transmitter to receiver which splits the message into small units or 'packets'. The splitting may be done either at the transmitting *terminal*, or at an *exchange*. Each packet includes the 'address' of the message's destination. Packet switching can then use different routes for the various parts of a message so as to make the most efficient use of the telecommunications network. Packets have to be sorted, monitored and reassembled at the reception point. There are two main techniques for packet switching: a. the *autonomous* mode where each packet is sent individually within the network according to the routing information attached to it; b. the *virtual link* mode where a path is pre-established within the network for each packet. See overleaf.

PAD packet assembler/disassembler. A device for attaching *terminals* which do not operate in a packet mode to a *packet-switched network*.

page 1. one side of a sheet of paper in a document (the commonest usage). 2. a section within a *computer memory*. 3. a term used in *viewdata* (interactive *videotex*). It represents an assembly of information in the *database* which can be accessed via its page number. A page may consist of several *frames*.

page printer a printer which sets the *character* pattern for a complete page before printing. It may be contrasted with a *line printer*.

page proof a printing term for the trial impression of a document laid out in the page divisions of the final product (see *proof*).

page reader an alphabetical character reader (see *optical character recognition*) which processes documents a page at a time.

page scrolling see *scrolling*.

page view terminal see *graphic display terminal (2.)*.

pagination 1. the (normally sequential) numbering of *pages* in a document. 2. a word-processing function used to create and, if desired, number *pages*.

paging 1. the division of *data*, or a *program*, into *pages (2.)* (ie sub-units). 2. the scanning of text on a *VDU* page by page, rather than by continuous *scrolling*. 3. as in radio paging, eg use of 'bleepers' etc to inform someone that they are wanted on, for instance, the telephone.

PAL Phase Alternate Line. This refers to one of two European standards for colour television broadcasting (the other is *SECAM*). It is also used in Australia and South Africa. (See also *NTSC*.)

palantype a *keyboard* which produces phonetic characters for subsequent transcription. (Transcription may be done automatically by computer.)

P

packet switched (or switching)

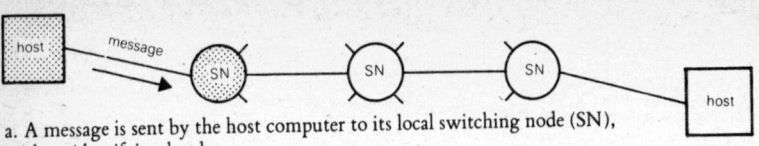

a. A message is sent by the host computer to its local switching node (SN), with an identifying header.

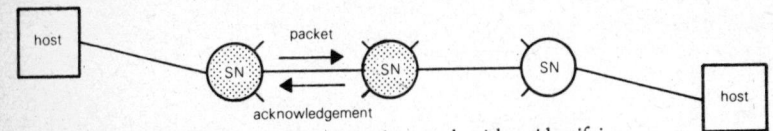

b. The local SN divides the message into packets, each with an identifying header, and sends them one at a time down the route to the next S/N.

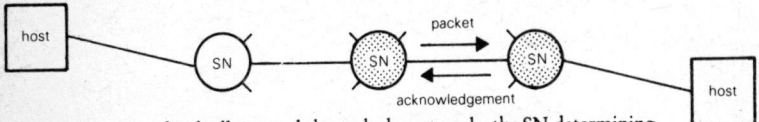

c. Each packet is individually routed through the network, the SN determining the routes and providing error detection and retransmission functions.

d. The destination SN assembles the message into its original form and passes it to the destination host computer.

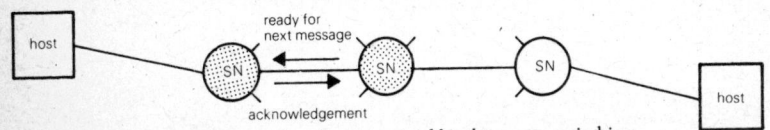

e. The destination SN sends a ready-for-next-message control packet to the source switching node, indicating that another message can be accepted.

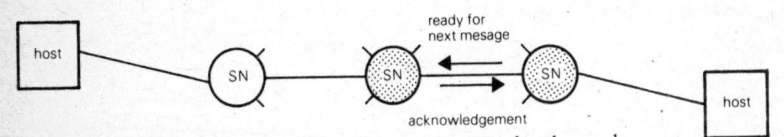

f. Only when the RFNM control packet is received by the source switching node can it accept another message from its host to the same destination.

Key:
SN = switching node
RFNM = ready for next message

Data flow in a packet switching network, (here ARPANET).

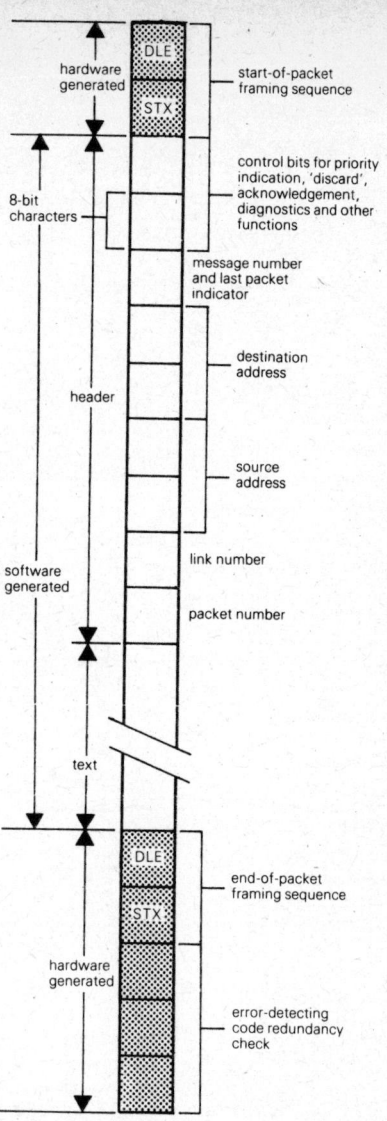

Structural diagram of a typical packet used in a packet switching network; in this case the ARPANET system. The text or data to be transmitted is preceded by a start-of-message signal and by bits for error-detection. This is followed by the header, which contains the destination and source address, the link number and the packet number. This ensures that packets are not lost or, if in error, are correctly transmitted. It is not necessary for the contents of the text to be investigated at any point, and there are usually safeguards against this.

pan see *scroll*.

PANDA *Prestel* Advanced *Network* Design *Architecture*.

paper-advance mechanism a device which drives paper through a *printer*. Typically, sprockets in the device engage with holes punched down each side of the paper.

paper tape used as a computer *input/output* medium, normally in reels a thousand feet in length and one inch wide. Information is recorded by means of punched holes: each *character* is registered as a row of holes across the tape. The choice of holes for each character is determined by the particular code being used.

paper tape reader a device which can detect the holes in a *paper tape* and translate them into *machine-readable* form.

parabolic dish an *antenna* whose cross-section is a parabola. Typically used in *satellite communications*.

parallel bit transmission a *data* transmission system where the *bits* representing a *character* are transmitted simultaneously.

parallel processing computer processing where more than one arithmetic operation is carried out at the same time (rather than sequentially as in most computers). Parallel processing finds particular use in some *pattern recognition* applications.

parallel-search storage see *associative storage*.

parallel transmission the simultaneous transmission of information either over distinct channels, or by different *carrier* frequencies over the same channel.

parent page a term used in *viewdata* (interactive *videotex*). A user is led to the *page* with the required information via *routing pages*. The routing page immediately prior to the required page is called the parent page.

parity bit a *binary digit* added to a set of *bits* so that the sum of all the bits is either always odd, or always even.

parity check a method of testing whether the number of ones (or zeros) in a set of *binary digits* is odd or even.

party line a communication channel which is shared, but without *multiplexing* (ie only one signal can be sent at a time). An example is the use of telephone party lines.

Pascal a *high level computer language*, particularly used for *systems programming*.

PASCAL Program Appliqué à la Selection et la Compilation Automatique de la Litterature. A very large multidisciplinary *database* compiled by the French Centre National de la Recherche Scientifique. PASCAL is accessible via *ESA/IRS*.

pass a complete computer processing run, including *input*, processing and *output*.

PASSIM President's Advisory Staff on Scientific Information Management, US.

password a group of *characters* which a user inputs to a computer to gain access to the system.

PATOLIS Patent On-line Information System. A system which allows *on-line searching* of Japanese patent information compiled by *JAPATIC*.

PATRICIA practical algorithm to receive information coded in alphanumeric (see *information retrieval techniques, algorithm* and *alphanumeric*).

pattern articulation unit a *microprocessor* which is used to reduce graphic images into *data streams*, and to reconstruct images from such data (see *character recognition, pattern recognition* and *OCR*).

pattern recognition (PR) a general term which can be used to cover recognition of any sort of stimulus, or input, eg visual, tactile, acoustic, chemical, electrical. It is normally used for the computer recognition of patterns (implying a reference to something already known) for the purpose of classification, grouping or identification.

The currently proposed main areas of application are in medical diagnosis, industrial inspection and control, and military systems. PR is essentially an information processing activity, which means that a range of techniques can be used. Those of general interest in information technology include *OCR* (optical character recognition), *MICR* (magnetic ink character recognition) and *speech recognition*.

PAU *pattern articulation unit*.

PAX *private automatic exchange*.

PBX *private branch exchange*.

PCM 1. *pulse code modulation*. 2. punchcard machine (see *card punch*).

PCMI photo-chromic micro image. An *ultrafiche* with up to 3200 images on an A6 size base. The production process requires a more stringently controlled environment than is necessary for ordinary *microfiche*.

PCS punched card system (see *punched card*).

peek a term used in the *computer language BASIC* when gaining access to *memory*.

Pel picture element (see *pixel*).

P/E News Petroleum/Energy News. A US *database* giving coverage of petroleum and energy business news. It is accessible via *SDC*.

PERA System a classification code for technical components developed by the UK Project Engineering Research Association (see *classification and coding systems*).

perfecting printing so that pages of text are printed on both sides and the paper is ready for folding.

perforated tape see *paper tape*.

perforator any device that can produce perforated *paper tape*.

Pergamon-Infoline a UK *host* giving access to a number of *databases*,

predominantly in the sciences.

peripheral see *peripheral unit*.

peripheral transfer moving *data* between two *peripheral units*.

peripheral (unit) a device under the control of a *central processor*. It may mean an *input* device, an *output* device, or a *storage* device.

permanent memory *memory* which retains its information when power is cut off. It is to be contrasted with *volatile memory*.

permutation index an *index* which lists all the words in a document's title so that each word appears, in turn, as the first word, followed by the remaining words.

permuted index see *permutation index*.

persistence the length of time for which a fluorescent screen (as used in a *CRT*) retains an image.

PERT programme evaluation and review technique. A management technique for scheduling and controlling large projects. It is more often called *critical path method* (CPM) or *network planning*.

PF 1. page footing. The record of information to be printed at the foot of a page, or pages, in a document. 2. pulse frequency.

PGI Unesco's General Information Programme.

PH page heading. The record of information to be printed at the head of a page, or pages, in a document.

phase encoding a method of recording *binary* digits on *magnetic tape* in which a '1' and a '0' are represented by different directions of magnetization. (See also *NRZI*.)

phase modulation see *modulation*.

Philips/MCA Discovision a *video disc* system developed by Philips and MCA. A master disc is created by exposing a surface to the action of a *laser* beam, which is *modulated* by the signals to be recorded. The laser beam creates pits in the surface, and these are reproduced in the copies made subsequently. Playback is achieved by focussing another laser beam onto the pitted surface. The reflected light is monitored and used to reproduce the original video (and audio) signals.

The discs, which have about 625 tracks per millimetre, are covered with a protective coating, and the reading laser is focussed on the surface below this. Consequently, the playback is unaffected by dirt, fingerprints, etc. There appears to be little limit to disc life, even when frozen on a single frame. This contrasts with videotape players, where use of a freeze frame facility can ultimately damage the tape.

It should be noted that MCA have now changed their name to DVA (Discovision Associates).

phoneme the smallest element of spoken language which distinguishes one utterance from another. For example, the word 'bit' consists of three phonemes – /b/, /i/ and /t/.

phosphor dot the element of a *cathode ray tube* which glows to form a *display*.

photocomposer often used as a synonym for *phototypesetter*.

photocomposition see *phototypesetting*.

photo-detector light sensitive device.

photoelectric detection detecting and reading marks with a *photoelectric detector*, as in *optical character recognition*.

photoelectric detector a device for detecting the presence of light and measuring its intensity.

photogravure a *gravure* printing process where the image is produced photographically.

photolithography a *lithographic* printing process where the image is formed photographically.

photonics another term for *fibre (fiber) optics* (see *optical fibre (fiber)*).

photo-optic memory *memory* which uses an optical storage medium (usually film).

photo-optic typesetter a *phototypesetter* in which the character master is a grid, or disc. Characters are optically projected from this onto a photosensitive surface.

photosensitive printing a general term for printing methods which depend on radiation, not on temperature-induced effects. These include *electrophotographic printing*, *photo-polymer printing* and some dye-based methods.

photosensor any light-sensitive device used in conjunction with an optical arrangement for *scanning* images, eg in *facsimile transmission*.

photosetting used to describe the typesetting of headlines by optical means, but is often employed as a synonym for *phototypesetting*.

phototelegram service *facsimile transmission* service operated by *British Telecom*. Graphic material is telegraphed from London world-wide over line circuits and radio links. The receiving office then mails the phototelegram to the final recipient by express, or registered, post.

phototelegraphy the name in Europe for newspicture transmission using *facsimile transmission*.

phototypesetter any device which makes it possible to set images of *type* on photographic material. (See also *computer-aided phototypesetting*.) Diagram on page 141.

phototypesetting the production of *type* images on a photographic medium by optical techniques. A machine that can perform this function is called a *phototypesetter*. (See also *computer-aided phototypesetting*.)

Physics Abstracts a *database* compiled by *INSPEC* in the UK, giving extensive bibliographic coverage of the physics literature. It is accessible via most of the major *hosts*.

pica see *point (system)*.

Pi character in *phototypesetting*, a *character* which is not normally stored on a *master grid* or in the computer's *memory*.

pico- a prefix meaning one million millionth (10^{-12}).

Picturephone a particular type of *video telephone*. 'Picturephone' is a trade name registered by the US company AT&T.

piece identification number see *PIN*.

PIN piece identification number: a machine-coded book label (see bar code).

pinfeed platen a *platen* which moves paper through a printer using 'pins'. These engage in holes along the edges of the paper.

PIRA Paper, Printing and Packaging Industries Research Association. A UK body which carries out research in all the subject areas suggested by its title (including *electronic publishing*). PIRA compiles a *database* (also called *PIRA*) in its areas of interest which is accessible via *Lockheed*.

PIRS personal information retrieval system. Refers to the storage of information on the *magnetic tapes* or *discs*, of a mini-

or *microcomputer*, using a *software package* to help order and index the information.

pitch number of characters per inch.

Pittler classification system a system for classifying technical components and describing them in *digital* code (see *classification and coding systems*).

pixel a picture element on a visual display screen (*VDU*). Diagram overleaf.

pixel pattern the matrix used in building the image of a character, or symbol, on a display screen.

plasma display a form of *display* which uses a flat panel as a screen. It has many advantages over a *cathode ray tube* display: most notably its compact size, good readability and internal *memory* capacity.

platen the plate in a printing machine against which paper is held in order to have printed material impressed upon it.

PL/1 Programming Language/1 a *high level programming language* with a wide range of scientific and business applications.

plotter a visual display device in which the values of one variable quantity are automatically plotted graphically against those of another.

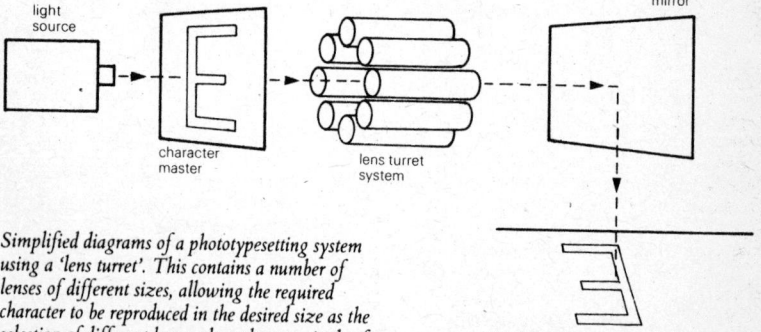

Simplified diagrams of a phototypesetting system using a 'lens turret'. This contains a number of lenses of different sizes, allowing the required character to be reproduced in the desired size as the selection of different lenses alters the magnitude of the image that is exposed on the photosensitive drum.

plug board

plug board a board into which electrical plugs can be inserted manually to control equipment.

plug compatible any two devices which will operate from the same socket are said to be 'plug compatible'.

PMBX see *private manual branch exchange*.

P MOS P-channel metal oxide semiconductor. A common form of *transistor* (where 'P' stands for positive – see *N MOS*).

PNI Pharmaceutical News Index. A US *database* covering journal articles on pharmaceuticals, drugs and cosmetics. PNI is accessible via *BRS*, *Lockheed* and *SDC*.

PO British Post Office: previously GPO (General Post Office).

POCS Patent Office Classification System, US.

point of sale describes electronic *terminals* used at retail outlets to record financial transactions as they occur (see *bar code*, *universal product code* and *British Telecom*).

point (system) a measurement of length in typography, used in English-speaking countries. One point = 0.351 mm; twelve points = one pica.

poke a term used in the *computer language BASIC* when putting information back into *memory*.

polarization the direction of vibration of an *electromagnetic wave*, and, correspondingly, the direction in which a receiving or transmitting *aerial* must be orientated.

polar transmission a method of transmitting *teletypewriter* signals which employs a three-state signal. (See also *neutral transmission*.)

POLIS Parliamentary On-line Information System. The *information retrieval system* of the House of Commons Library in the United Kingdom.

polling a means of controlling communication along a series of *lines* by seeing if any of them is waiting to deliver a message.

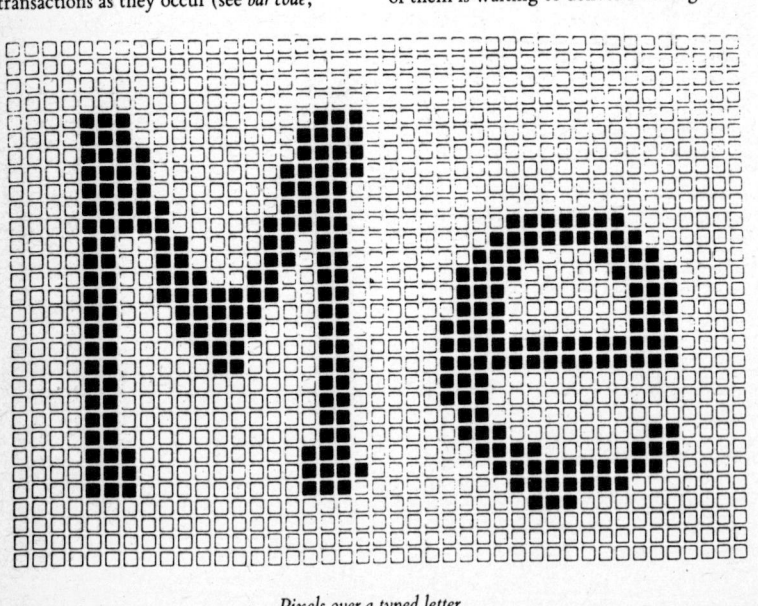

Pixels over a typed letter.

Pollution Abstracts a US *database* covering pollution, environmental quality, pesticides and related topics. It is accessible via *BRS, ESA/IRS, Lockheed, QL* and *SDC*.

port a place of entry to, or exit from, a *central processor*.

POS *point of sale*.

POSH permuted on subject headings (see *permuted index*).

positive an image (usually photographic) with *tones* which are the same as those of the original.

post coordinate indexing see *indexing*.

post editing carrying out *editing* on *output*.

postings in *information retrieval*, the number of records *retrieved* by a *search*.

PPS pulses per second. A unit of signal transmission rate.

precedence used as a synonym for *priority*.

precedence code a *code* which signifies that the *characters* in the following code (or codes) will have a different meaning from normal.

PRECIS 1. preserved context index system. A subject *indexing system*, developed for the British National Bibliography, in which initial *strings* of *terms* are organized according to their linguistic function. The computer manipulates the strings so that selected words function in turn as the 'approach' term (ie the term in which the user is primarily interested). 2. pre-coordinate indexing system. A system in which terms are combined at the time of indexing a document. The combination of terms is recorded in the entries (see *indexing*).

precision 1. a measure of exactness (for example, the number of decimal places to which a value is shown). Should not be confused with *accuracy*. 2. in bibliographic *information retrieval systems*, 'precision' means the percentage of all *items* retrieved by a search on a particular topic which

actually prove to be relevant to that topic. (See also *recall*.)

precoordinate indexing see *indexing*.

Predicasts Files these contain over three million summaries of information from documents covering a variety of business and industrial statistics. Most of the files are accessible for *on-line searching* via *Lockheed*.

pre-editing carrying out *editing* prior to *input*.

Prestel the *viewdata* (interactive *videotex*) system implemented by British Telecom in the UK. Diagram overleaf.

Pricedata a *databank* covering world commodity prices. It is compiled in Italy, and is accessible via the *ESA/IRS host* computer and *Euronet-Diane*.

printed circuit board a mass-produced unit consisting of electrical components on a board and interconnected to produce a *circuit*.

printer in new information technology, this refers to an *output* device which converts electronic signals into print-on-paper (see *line printer, page printer, daisy wheel printer*).

printerfacing the provision of an interface between *microcomputers* and output printing *terminals*.

printer limited when the relatively low speed of the *printer* is the limiting factor in determining the rate at which *data processing* can take place.

printer plotter a printer (usually of the *daisy wheel*, or *matrix* type) which can undertake graphics reproduction in addition to character printing.

printer spacing chart a form used in deciding the format of printed output.

printer's type the range of different *characters* and *typeface* designs available from a printing office.

printout the printed paper *output* (pages, or continuous roll) which a computer produces via a *printer*.

print wheel a wheel (which can be changed when necessary) used to print characters in some types of *printer* (see eg *daisy wheel printer*).

priority refers to the transmission of messages following some designated order of importance.

PRISM Personal Records Information System for Management. A computerized personal information system used by the civil service in the United Kingdom.

private automatic exchange (PAX) an exchange for a private telephone service, within an organization, which is not connected to the public telephone network.

private branch exchange (PBX) a *switching* facility (exchange) within an institution which also provides access to the public telephone network. The exchange may be manually operated, in which case it is, strictly, a private manual branch exchange (PMBX), or, it may have an automatic switching facility, and so be classified as a private automatic branch exchange (PABX).

private line a channel with associated equipment provided for the exclusive

Prestel: this diagram illustrates the three types of user terminal available; (a) a domestic television set with viewdata, *(b) a business terminal and (c) an information provider's terminal.*

use of a particular subscriber.

problem-orientated language a *high level programming language* whose structure depends on the specific nature of the problem with which it is designed to deal.

process colo(u)r work the equivalent of four-colo(u)r reproduction (ie the use of combinations of red, yellow, blue and black to give a full range of colo(u)rs).

processor synonymous with *central processor.*

profile a term used for the range of interests of an individual or group in *SDI.*

program an ordered list of *instructions* directing a *computer* to carry out a desired sequence of *operations.* The objective is normally the solution of a problem.

program crash a *program* crash occurs when a computer program attempts to execute an impossible instruction, but has no way of recognising the impossibility and stopping. Well-written programs which are unlikely to crash are referred to as 'crash-proof' or 'robust'.

program flowchart a *flowchart* diagram which describes a computer *program* in terms of a series of steps, and of the relationship between these steps. Standard sets of symbols are used in constructing the flowchart. The most common such set is illustrated overleaf.

program library see *library.*

programmed learning see *computer assisted instruction* and *educational technology.*

programmer a person who writes *programs.*

programming language see *high level programming language.*

program specification a description of a *program* which specifies all the information needed to design it in detail. The description normally includes a *program flowchart.* (Compare with *documentation.*)

PROM programmable read-only memory. (See *ROM.* See also *blowing.*)

prom blowing see *blowing.*

PROMIS Problem Orientated Medical Information System. A very flexible computer-based medical record and hospital information system developed in the US. All data are entered and retrieved electronically, using *terminals* with *touch-sensitive screens.*

prompt a message given to an *operator (1.)* by an *operating system.* It usually indicates that particular information is required before a *program* can proceed.

Pronto the trade-name of a five-finger *keyboard,* having eight *keys:* three shift keys and five (one for each finger) character and symbol selection keys. It has the advantage over conventional keyboards of being small and light (potentially portable) and allowing data entry to be performed with one hand.

proof in printing, a trial impression, obtained after *composition,* which can be used to introduce corrections. (See also *galley proof* and *page proof.*)

proportional spacing in *typesetting,* the horizontal spacing of characters in proportion to their width. This may be contrasted with the constant spacing used on a standard typewriter.

proprietary software copyright *software* sold on a commercial basis.

protected location a computer *location* in which data cannot be stored without undergoing some test procedure beforehand.

protocol a set of conventions governing the *format* of messages to be exchanged within a communication system.

PROXI protection by reflection optics of xerographic images (see *xerography*).

PRR pulse repetition rate. The number of electronic pulses received in unit time at a

symbol	represents

PROCESSING
A group of program instructions that perform a processing function within a program.

INPUT/OUTPUT
Any function of an input/output device (making information available for processing, recording processing information, tape positioning, etc.).

DECISION
The decision function is used to document points in a program where a branch to alternate paths is possible based upon variable conditions.

PROGRAM MODIFICATION
An instruction or group of instructions which changes a program sequence.

PREDEFINED PROCESS
This identifies a group of operations not detailed in a particular set of flowcharts.

TERMINAL
The beginning, end, or a point of interruption in a program.

CONNECTOR
An entry from, or an exit to, another part of a program flowchart.

OFFPAGE CONNECTOR
A connector used to designate entry to or exit from a page.

FLOW DIRECTION
The direction of processing or data flow.

ANNOTATION
The addition of descriptive comments or explanatory notes as clarification.

Selection of symbols used in program flowcharts.

specified point in a *computer*.

PSS packet switching service (see *packet switching*).

PSU power supply unit.

Psychological Abstracts *bibliographic database* covering psychology and related subject areas. It is accessible via *BRS*, *DIMDI*, *Lockheed* and *SDC*.

PTS F and S Indexes one of the biggest business information *databanks*. It contains information on companies, products and industries, and includes sales figures and profit forecasts. Based in the US, it has international coverage and is accessible via *Lockheed*.

PTS Files Predicasts Terminal System Files (see *Predicasts Files*).

PTT Postal, Telegraph and Telephone Authority.

pulse code modulation (PCM) in pulse code *modulation*, the wave form of the signal to be transmitted is sampled at fixed time intervals. Its magnitude at each interval is transmitted in the form of *digital* pulses, so as to facilitate high speed transmission. (See also *analog(ue) transmission* and *digital transmission*.)

punch 1. to make holes in a *card*, or *paper tape*, so as to enter information. 2. a piece of equipment to make such holes.

punched card see *card*.

punched tape see *paper tape*.

pure machine-aided translation an approach to the design of *machine translation*. The lexicons of two, or more, lan-

guages are computerized in order to supply a human translator with *target language* equivalents of *source language* lexical items. Usually only specialized vocabulary is included in an *automated lexicon*, leaving the common core vocabulary to the human translator (see *machine-aided translation* and *machine translation*).

pure machine translation a *machine translation* system in which the computer attempts to do the entire translation itself. Only pre-input and/or post-output text editing by human translators is required. Despite a great deal of research and development, pure machine translation is still in the experimental stage, while *machine-aided translation* has become an operational reality (see *HAMT* and *SYSTRAN*).

pure MAT *pure machine-aided translation*.

pure MT *pure machine translation*.

purge to erase *data* from a *file*.

pushbutton dialling the use of pushbuttons instead of a rotary dial for feeding in a sequence of numbers or letters, eg to a telephone. Each number/letter may be identified by its own audio signal.

pushdown list a list of items compiled in such a manner that the last item added stands at the top of the list. (See also *pushup list*.)

pushup list a list of items compiled in such a manner that the last item added goes to the bottom of the list. (See also *pushdown list*.)

PVT page view terminal (see *graphic display terminal (2.)*).

PW private wire (see *private line*).

QAM *queued access method.*

Q-band *frequency band* used in radar (36-46 GHz).

QED a *software package* for *text editing*.

QL Quick Law Systems. A Canadian *host* offering access to *databases* containing legal information.

QTAM *queued telecommunications access method.*

quadding a term used in typesetting. It refers to the insertion of blank space, or to the blank space itself, in a typeset line.

quadruplex transverse scanning see *transverse scanning.*

quantizing error distortion brought about by *analog(ue) to digital conversion*. It occurs when *analog(ue)* signals fall between the possible *digital* values (see *analog(ue)-to-digital conversion*).

Quebec-Actualité Canadian *database* covering items of Canadian and world news drawn from several French-Canadian newspapers. It is accessible via *SDC*.

query language a *high level programming language*, resembling a *natural language*, which is designed to make *on-line searching* easy for inexperienced users.

Quest a computer language used for searching *databases* via *ESA/IRS*.

Questel a French *host* covering a number of French and EEC *databases* in science, technology and business.

queue a set of *jobs* awaiting processing.

queued access method a *data processing* method in which the transfer of *data* between devices is automatically synchronized to eliminate delays.

queued telecommunications access method when the transfer of *data* between a computer and its *peripherals* is automatically synchronized via *telecommunications channels* (see *queued access method*).

quick access memory *memory* with relatively short *access time*.

QUICKTRAN Quick FORTRAN. A *high level programming language*, developed from *FORTRAN*, designed for use in *conversational mode* on *time sharing* systems.

Qwerty keyboard a *keyboard* which has the keys laid out in the same pattern as that used on ordinary typewriters. The word derives from the letters at the top left of the keyboard.

Q

RACE random access computer equipment (see *random access*).

RAD rapid access disc (see *magnetic disc* and *access time*).

RADA random access discrete address. A location in a *RAM* (see *address*).

radio communication any communication using radio waves (see *spectrum*).

Radio Suisse a Swiss *host* system.

RADIR random access document indexing and retrieval (see *random access*, *RAM*, and *information retrieval system*).

ragged right an uneven right-hand margin (see *justify*).

RAM *random access memory*.

RAMIS Random Access Management Information System. An *information retrieval system* which stores management information in *RAM*.

RAMPI Raw Material Price Index. An *online databank*.

R&D research and development. Designates technical and applied scientific activity, particularly directed towards the development of new products, processes, services or systems.

random access memory (RAM) *memory* where any location can be read from, or written to, in a *random access* fashion.

random access (storage) *access* to *storage* where the next *location* from which information is to be obtained is unrelated to the previous location. Normally implies that the *access time* to any location is the same.

ranged left text which is justified at the left-hand margin only. Used as a synonym for *ragged right* (see *justify*).

RAPID random access personnel information system (see *RAMIS*, *RAM* and *information retrieval system*).

RASTAC random access storage and control (see *RAM*).

RASTAD random access storage and display (see *RAM* and *display*).

raster a grid on a *terminal* screen which divides the *display* area into discrete elements (like a map reference system).

raster count the number of positions on a *display* screen which can be defined using its *raster* (ie the product of the number of horizontal and vertical divisions).

raster graphics a form of *computer graphics* which, unlike *vector graphics*, utilizes a full matrix of *pixels*. Each pixel has its own code, and is switched on, or off, according to a guiding *program* (see *raster*).

raster plotter a *plotter* which draws a complete picture on a *CRT*, including an image both of the object of interest and its background. It is used in *computer graphics*. (See also *calligraphic plotter* and *raster*.)

raster scan the sweeping of the display area of a device, line-by-line, to generate, or read, an image.

raw data *data* which have not been processed.

RAX *remote access*.

RCA Selectavision see *Selectavision*.

reactive mode when each entry at a *terminal* causes some action to be taken by a *central processor*, but the processor does not necessarily return an immediate response to the *terminal*. It is to be contrasted with *conversational mode*.

read 1. to copy, usually from one *storage* area to another. 2. to sense information from some form of recorded medium, eg from a *card* or *magnetic tape*.

READ 1. real-time electronic access and display (see *real time*, *access* and *display*). 2. remote electrical alphanumeric display (see *remote access*, *alphanumeric* and *display*).

R

read-only memory (ROM) once information has been entered into this memory, it can be read as often as required, but cannot normally be changed. Currently available *video discs* are read-only devices.

readout *soft copy* output from a computer displayed on the screen of a *VDT*.

read screen a transparent screen through which documents are read in *optical character recognition*.

read/write head an *electromagnetic* device used to *read* from, or *write* on, a *magnetic storage device*: similar to a 'pick-up' head on an ordinary audio tape recorder.

read/write memory information written into a computer *memory* or *storage* area can then be accessed and read. In a read/write memory (or store) this information can be altered at will and read again as often as required. This may be contrasted with *read-only memory*.

read/write storage see *read/write memory*.

ready a *status word* indicating that a *computer* is waiting for *input* from a *terminal*.

REALCOM Real-time Communications: a system developed by RCA (see *real time*).

real time the computer response occurs at the same rate as the data input. Most control systems, eg an automatic pilot on an airplane, operate in real time, as do *on-line search* systems.

recall in bibliographic information *retrieval systems*, recall is the ratio of all items actually retrieved by a search to the total number of items searched for. It is usually expressed as a percentage. (See also *precision*.)

recognition logic the *software* in an *OCR* reader which allows it to translate printed text into *digital* form.

recognition unit a device for interpreting the electrical signals received when *scanning* a document in order to capture text in *machine-readable* form. It matches the 'read'

character against its own store of characters in order to 'recognize' it for conversion into computer code. (See also *optical character recognition*.)

RECOL Retrieval Command Language. A *computer language* used for the interrogation of *databases* within an *information retrieval system*. (See also *query language* and *search language*.)

record a unit, or set of data, forming the basic element of a *file*.

record separator a *character* indicating the boundary between two *records* in a *file*, or transmission.

recovery system a computer *program* which records the progress of computer *processing* activities, to allow reconstruction of a *run* in the event of a computer *crash*.

reduction rates in *micrographics*, this refers to the ratio between the scale of the original material and the scale of the *microform* image.

redundancy 1. the proportion of information in a message which can be eliminated without the message losing its essential meaning. 2. provision of a back-up system which can be automatically activated if the primary system breaks down.

redundancy check a method of checking for the presence of errors in *data*. It depends on the use of more bits than are required to represent the information concerned. The check is usually carried out automatically (see, eg *parity check*).

refereed a document which has been subjected to evaluation by experts and passed as suitable for publication.

reference see *citation*.

refresh the process of reactivating information. This is required, for instance, with a *VDU* display. Material appearing on the screen is generated by the action of cathode rays on phosphor dots. The phosphors light up, but then immediately begin to fade until reactivated. Refreshing tends to create flicker. (See also *storage tube*.)

refresh rate the rate at which a *CRT* screen is refreshed (see refresh). A minimum refresh rate of 50 Hz (cycles per second) is recommended for comfortable viewing.

regenerative memory *memory* which needs to be *refreshed*: otherwise the contents disappear.

register 1. a computer *storage* device which holds data, *addresses* or instructions on a temporary basis. For example, if a series of numbers is being added, the intermediate totals can be accumulated in a register as they are produced. 2. obtaining the correct position relative to each other of two or more printings on the same sheet.

relational database a *database* in which the relations between the *items* it contains are formally stated.

relay an electrically or electronically operated switch.

Relay a US *communication satellite* launched in the 1970s.

remote access the use of a *computer* from a *terminal* at a geographically distant point. The terminal and computer may be connected via either cables, or broadcast transmission.

remote batch processing the sending of *data* and *programs* in *batches* to a *central processor*.

remote batch terminal *terminal* used for *remote batch processing*.

remote job entry (RJE) entry of data, or operating instructions, into a computer from a *remote terminal*. (See also *job*.)

remote printing the production of *hard copy output* from a *printer* situated in a geographically distant location from the *processor* which provides the printer's electronic *input*.

remote terminal a *terminal* which is in a

geographically distinct location from the *processor* which it is accessing.

REMSTAR remote electronic microfilm storage transmission and retrieval (see *COM*).

repeater 1. a device for restoring signals, which have been distorted through *attenuation*, to their original shape and transmission level. 2. a device whereby currents received over one *circuit* are automatically repeated over another circuit, or circuits.

resolution the fineness of detail that can be distinguished in an image. (Often expressed in terms of number of distinguishable lines per cm.)

response frame this is a *viewdata* (interactive *videotex*) term. It refers to a *frame* which expects a response from the user. The response is communicated to the *information provider*.

response time the time interval between an event and the system's response to the event. For example, in computers it might be the time between the pressing of the last key when inputting at a *terminal*, and the *terminal's* display of the first character of the response.

retrieval see *information retrieval systems* and *information retrieval techniques*.

retrospective searching an *end-user* makes a *search request* to a *database* in the form of a call for all items published on a specific topic since a specified date. The user's search request is converted into *search terms*, and then translated into *machine-readable* language. All items in the database which have been indexed under these terms are identified in the course of a single run. The search may be made either *interactively*, or as part of a *batch*.
Retrospective batch searching is cheaper than *on-line searching* for the user, but has less flexibility (see *information retrieval systems*).

reverse channel capability the ability to interact with a system over a communications link.

reverse leading movement of the film (or paper) in a *phototypesetter* in the opposite direction to normal operation.

revise any proof produced after corrections have been made.

rewind return a *magnetic tape* to its beginning.

REX real time executive routine. A computer *routine* which is to be executed in *real time*.

RF radio *frequency* (see *spectrum*).

RF modulator radio frequency modulator. A device used to *modulate* the *frequency* of a *carrier signal*. It is frequently used to convert the *output* signal from a *microcomputer* into a form which can be displayed on a normal television screen.

RICASIP Research Information Center and Advisory Service on Information Processing. A body, whose aims are indicated by its title, jointly sponsored by the National Science Foundation and the National Bureau of Standards in the US.

right justify sometimes used as the equivalent of *flush right*.

rigid disc see *hard disc*.

ring code used for Pharmadokumentationring, which comprises a group of European pharmaceutical companies who have collaborated in indexing material for *chemical structure retrieval*.

Ringdoc a UK *bibliographic database* covering pharmaceuticals. It is accessible via *Pergamon-Infoline* and *SDC*.

ring network a network arranged in a way that permits terminals to communicate with one another without having to go via a central computer. (See also *local area network*.)

rivers an undesirable alignment of spaces in a text.

RJE *remote job entry*.

RO receive only – equipment which can receive, but not transmit.

robot a machine capable of automatically performing some type of activity which is normally controlled by human beings. Most existing robots carry out simple repetitive tasks; most commonly in an industrial setting, as on an automatic production line. Robots are often linked to a *numerical control*, or *computer numerical control* system. (See also *artificial intelligence*.)

robotics the application of *artificial intelligence* techniques to the design and production of *robots*.

role indicator in *information retrieval*, a *code* assigned to a word, eg a *descriptor* or *keyword*, which indicates the role, eg part of speech, which the word plays in the text where it occurs.

ROM *read-only memory*.

Roman commonest form of *type face*.

routine a sequence of operations for a computer to perform.

routing the assignment of the communications channel by which a message can reach its destination.

routing indicator an *address*, or group of *characters*, at the beginning of a message which indicate its final destination.

routing page this is a *viewdata* (interactive *videotex*) term. A *page* whose function is to indicate a choice of other pages.

RPG Report Program Generator. A *high level programming language* used in drawing up business reports.

RPM revolutions per minute.

RRP reader and reader-printer. A device for viewing *microforms* and producing *hard copy* from the microform as required.

RS *record separator*.

RT 1. real time. 2. related term: a cross

reference in a *thesaurus*. 3. remote terminal: a *terminal* with *remote access*.

RTECS Registry of Toxic Effects of Chemical Substances. A *database* compiled by the US National Institute for Occupational Safety and Health.

RTU remote terminal unit. A computer *terminal* with *remote access*.

run one execution of a computer *routine*, *program* or *suite* of programs.

running head a line of *characters* at the top of each page of a document which provides information concerning the document, eg author, title, chapter.

R/W *read/write*.

R/W memory (or storage) see *read/write memory* (or *storage*).

s second: *SI* unit of time.

S siemen: *SI* unit of conductance.

SALINET Satellite Library Information Network. A *satellite communications* system used to provide library services to remote parts of Canada.

SAM *serial access memory.*

Samson a Dutch *host* offering *databases* via *Euronet-Diane*. Specializes in maritime information.

sans serif a *typeface* without *serifs*.

Satcom a series of *communications satellites* owned by RCA for communication in the US.

Satellite Business Systems established by Comsat General (a subsidiary of *Comsat*), with *IBM* as the dominant partner, to offer *satellite communication* channels within the US. One of its intentions is to provide a US *electronic mail* service.

satellite communication communication via satellites offers advantages over both radio and cable transmission. The main problems with ground-based radio transmission are: interference between different transmissions; atmospheric *attenuation*; and the propagation of radio waves in straight lines. Satellites help overcome these problems because: i. the *antenna* beam covers only a limited area of the Earth's surface; ii. attenuation is reduced since the radio waves pass at a relatively steep angle through the atmosphere; iii. the altitude of the satellite means that the radio waves are little affected by topographical obstacles. As compared with international communication via cables, satellite links are more flexible and cheaper (currently a third or less per channel). Diagram overleaf.
Most communications satellites are placed in a *geostationary orbit*, since then ground-based antennae can be permanently turned to the same point in the sky. Such an orbit has the disadvantage that the satellite is fairly low in the sky for receivers in high northern and southern latitudes. More importantly, the number of satellites that

can be placed in a geostationary orbit is ultimately limited, along with the *frequencies* at which they can receive and retransmit signals. For this reason, successive *World Administrative Radio Conferences* (WARCs) have tried, with moderate success, to impose international regulations on the use of communication satellites. Satellites are already in use for telephone, data, radio and television transmissions. Frequencies in the *gigahertz* range are employed, normally with antennae of ten metres, or more, diameter. Plans are currently in hand to increase the power of satellite transmissions, and so decrease the size of the ground-based antenna required. This might permit the introduction of home-based antennae; for example, to receive television transmissions directly (so competing with *cable television*). At present the *space segment* consists primarily of a *transponder* which amplifies and retransmits the incoming signal (usually with frequency conversion). *Switching* activities are controlled by the *ground-based segment*.
Much the largest satellite communication system is *Intelsat* (which has the US body, *Comsat*, as its major shareholder). Individual US firms have also placed communications satellites in orbit, eg *Satcom* and *Westar*, and a marine version of Intelsat (called *Inmarsat*) has been sent up. *ESA* and the Soviet Union have already established major program(me)s in this field. The former has experimented with *OTS*, which is planned to lead on to an operational series of *ECSs*. The Soviet Union has launched a large number of *Molinya* communications satellites (which are not geostationary), and now plans to launch a series of *Statsionar* geostationary satellites.
Most of these satellites can be used for any type of communications traffic, though they may, for organizational reasons, be dedicated to one type of transmission only.

satellite computer a smaller ancillary computer used to relieve a central, larger computer of relatively simple, but time-consuming operations.

Sat Stream an international *digital data transmission* service planned by *British Telecom* for introduction in the 1980s. It will use the *Switchstream network* in the UK,

S

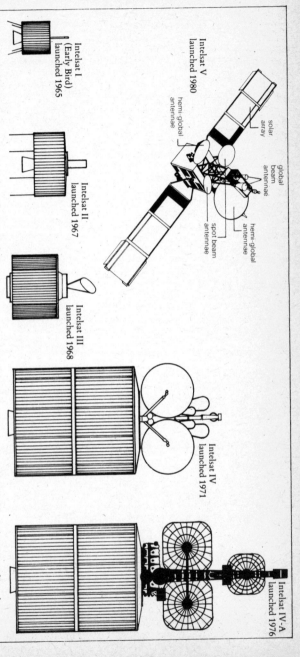

Intelsat I
(Early Bird)
launched 1965

Intelsat II
launched 1967

Intelsat III
launched 1968

Intelsat IV
launched 1971

Intelsat IV-A
launched 1976

Intelsat V
launched 1980

solar
array

global
beam
antennae

hemi-global
antennae

hemi-global
antennae

spot beam
antennae

The range of INTELSAT communications satellites. Intelsat I had a capacity of 240 telephone circuits or one TV channel and was intended purely for point-to-point communications between Western Europe and North America. Intelsat II had the same capacity but could be accessed by more than just two earth stations. Intelsat III offered 1500 circuits or four TV channels. Intelsat IV no fewer than 4000 circuits or twelve TV channels. It had not only multiple access but also simultaneous transmission capabilities. Intelsat IV-A provides 6000 circuits and two TV channels, or twenty TV channels if devoted entirely to that medium. Intelsat V (above left) is of radically different shape to its predecessors and is capable of carrying 12000 circuits and two colour TV channels, this being made possible by using two frequency bands (6/4 GHZ and 14/11 GHz). It is the largest commercial satellite currently in operation.

and beam messages via two satellites run by *ESA* and the French telecommunications administration.

SBS *Satellite Business Systems*.

scan in information technology, this means to examine material, eg a page of text, the data present typically being converted into *machine-readable* form (see, eg *optical character recognition*).

scanner a device for examining printed characters, or *graphics*, and representing them by electrical signals. Used especially for devices which produce a *digital* output which can be input to a computer.

SCANNET Scandinavian Network. A Swedish *host* and *computer network*.

scanning the sequential examination, or exposure, of a set of *characters*, or of an image (see *facsimile transmission* and *optical character recognition*).

scanning device a general term for a *scanner*, but also has the specific meaning of an attachment to a *microform reader*. It allows the user to bring any section of the microform to a position in which it can be most easily read. (See also *rapid scan system*.)

SCI *Science Citation Index*.

SCIM Selected Categories in Microfiche. An *SDM* service operated by the US *National Technical Information Service*.

SCI Search the *database* of the Science Citation Index, produced by the Institute of Scientific Information in the US. *On-line searching* of the database can be made via *Lockheed* and *DIMDI*.

screen 1. a surface (especially of a *CRT*) where information can be displayed. 2. the method of forming dots in *half-tone* illustrations.

screenload the maximum number of *characters* that can appear on a *screen* at one time.

scroll movement of text up and down, or

across, a *VDU* so that the user can view areas of text adjacent to that displayed on the screen. It is used, for example, on some *word processors* where the screen will not hold all the contents of an A4 page. Scrolling across a VDU is sometimes referred to as 'panning'.

SDA *source data acquisition*.

SDC System Development Corporation. One of the largest database *hosts* in the world, it offers *on-line* access to over 50 *databases* covering a wide range of topics.

SDI *selective dissemination of information*.

SDILINE Selective Dissemination of Information On-line. A *current awareness service* giving access to the most recent records added to the *MEDLINE database*. It is accessible via *Blaise* or *NLM*.

SDM Selective Dissemination on Microfiche. A service which provides subscribers with copies of documents, in *microfiche*, in their (pre-specified) areas of interest.

search a systematic examination of information in a *database*. The aim is to identify *items* which satisfy particular pre-set criteria (see *information retrieval system* and *on-line searching*).

search language term used to describe any language used in the search of a *database* (see *information retrieval system* and *techniques*, and *on-line searching*). Also called a *command language* or *query language*.

search terms words, or groups of words (often *keywords*), used in *on-line searching* when specifying a request for information. Search terms correspond to the headings under which items in a *database* are indexed.

search time the amount of time required to locate a particular *item*, or *field* of *data*, in a store.

Secam the name stems from the French: 'Séquential à Mémoire'. It is a colo(u)r television system used in France, the USSR and Eastern Europe, together with some African and Middle Eastern countries.

Unlike the *PAL* and *NTSC* systems, where the two colo(u)r-difference signals are transmitted simultaneously, in Secam they are transmitted alternately.

sector the smallest unit of *memory* on a *magnetic disc* or *drum* which can be separately *addressed*. The term is also sometimes used to refer to a *block* of data occupying such a unit (see *soft sectoring* and *hard sectoring*).

Selectavision a *video disc* system developed by RCA. Grooves are produced in a master disc using a piezoelectric *stylus*. Piezoelectric crystals convert electrical signals into mechanical motion, and vice versa. The resultant disc has about 400 grooves per millimetre, and it is used to provide plastic copies.

Selected research in microfiche a *selective dissemination of information* service which not only provides references to new documents, but also delivers these documents on *microfiche*.

selective calling occurs when a transmitting station can specify which of a number of stations on the same *line* is to receive a particular message.

selective dissemination of information a general term for the provision of a *current awareness service* based on a *bibliographic database*. A subscriber to the service provides a profile of his/her interests (a *user interest profile*), which is translated into *machine-readable* form, stored on *magnetic tape*, and then matched at regular intervals against new additions to the *database*. Any additions matching the subscribers' interest profile are printed out and supplied to the subscriber. (See also *Group SDI*.)

Selective Dissemination on Microfiche see *SDM*.

self-checking code see *error-detection code*.

SEMCOR Semantic Correlation. A computer-aided *indexing system* (see *automatic indexing*).

semiconductor any substance which conducts electricity easily when the voltage across it is above a certain value, but not when it falls below that value. Semiconducting materials form the basis of *transistors*.

semiconductor laser a small *laser* made from semiconducting material, eg gallium arsenide. They can be used to emit beams of light down *optical fibres* for telecommunications purposes.

semi micro xerography see *SMX*.

sentinel sometimes used as a synonym for *flag*.

sequential access a computer term meaning that access to *memory* follows a prescribed order.

serial to handle items, or actions, sequentially.

serial access memory *memory* in which data are entered in sequence. This leads to sequential *processing* of the data (see *serial processing*).

serial bit transmission a *data* transmission system in which the *bits* representing a *character* are transmitted consecutively.

serial file a *file* in which *items* are entered in sequence so that they must also be searched sequentially. No provision is made for selective searching (see *information retrieval systems*).

serial printer a computer *printer* which prints one character at a time along a line (rather than a whole line together). An example is the *daisy wheel printer*.

serial processing processing items in a data *file* in the order in which they are *stored*.

serial storage *storage* in which *words* appear in sequence. This means that *access time* will include a waiting time.

serif a typeface in which the ascending and descending parts of the characters have small extensions.

SERLINE Serials On-line. The catalog(ue) of journal titles received by the US *National Library of Medicine*, which can be made available *on-line*.

service bureau a facility, usually commercial, which allows a computer user to lease time on a *central processor* and appropriate *peripherals* to run his/her *programs*.

set width measures the width of a piece of type.

shared logic a computer system in which *logic* (or *intelligence*) is shared between items of *hardware*, eg some *word processing* systems which operate under central control.

shared resource refers to a system in which the units make common use of a particular facility, eg *storage*, *printer*. Includes shared *logic*, shared storage, and cluster systems. Frequently occurs in *word processing*.

sheet microfilm a sheet of film containing frames of microphotographs in a rectangular pattern.

shelf life the length of time a document, or device, will remain of value to users.

SHF super high frequency (between 3000 and 30,000 mega Hertz).

shift movement of characters to the left, or right, by a prescribed number of places.

shift down modem a *modem* which produces a change from a higher to a lower *bit rate*.

SI Système International. A metric system of measurement units, based on the metre (length), gramme (weight), second (time) and ampère (electrical current). SI units are currently superseding the *British Imperial System* (in the UK) and the *US customary system* (in the US).

SIGLE System for Information on Grey Literature in Europe. A project sponsored by the Commission of the European Communities to improve the detection, identification, collection and delivery of 'grey' literature.

Grey literature broadly refers to material not formally published, although sometimes made widely available. (The main form is technical reports.) The initial project aims to set up an *on-line bibliographic database* to be made available through *Euronet-Diane*.

signal the expression of information in the form of electrical disturbances. Also the act of transmitting such disturbances.

signal-to-noise ratio ratio of the power of a *signal* to the *noise* in a communications *channel*. The higher the ratio, the easier the signal is to detect.

silicon chip a wafer of silicon providing a *semi-conductor* base for a number of electrical *circuits* (see *chip*).

simplex when transmission can only be carried out in one direction.

SIR selective information retrieval (see *information retrieval system*).

skew to be incorrectly aligned, eg a skew picture in *facsimile transmission*.

skew character a form of incorrect registration in *optical character recognition*.

skew failure when a *document* in *machine-readable form* cannot be *read* because it is not aligned properly in the reading device.

skipping advancing paper through a *printer* without printing upon it.

slab in computing, a part of a *word*.

slanted abstract an *abstract* which emphasizes a particular aspect of a document, in order to cater for the interests of a particular user group.

slave a device which operates under the control of another device. Especially a device driven by a computer's output.

slave tube a *CRT* connected to another CRT in such a way that each gives an identical display.

sloped Roman see *oblique*.

small cap(ital)s capital letters of a similar size to ordinary *lower case* letters.

SMART 1. System for the Mechanical Analysis and Retrieval of Text. (An occasional variant is Salton's Magical Automatic Retrieval Technique.) A system designed for the *interactive* search of full text documents. A user *inputs search words* and groups of words (phrases, etc). SMART analyses the text and produces lists of documents ranked in terms of the relative frequency with which the search words and phrases appear in each document (see *information retrieval techniques, natural language searching*).
2. a *Pure-MAT* system which uses an *automated glossary* to provide translation between Arabic, English, French, German and Spanish. It has an exceptionally large *database* covering engineering and military technology (see *pure-MAT* and *machine-aided translation*).

smart terminal a *terminal* which has some *data processing* capability of its own, but not as much as an *intelligent terminal*.

SMX semi micro xerography. A method of *xerographic* reproduction. It produces copies of documents which can then be input in *micrographic* form, via a special *reader*, to a computer.

S/N *signal-to-noise ratio*.

sniffing an error detection and correction method in computing.

SNOBOL a *high level programming language* designed for the *processing* of *character strings*.

SNR *signal-to-noise ratio*.

SOCIAL SCISEARCH Social Science Citation Index Search. The *Social Science Citation Index* on *machine-readable files*, which can be searched via *BRS* or *Lockheed*.

Sociological Abstracts a *bibliographic database* covering sociology and related fields. It is accessible via *BRS* and *Lockheed*.

soft copy 1. computer *output* displayed on a *VDU*. 2. computer output in some medium which cannot be directly read by the unaided eye, eg *floppy disc*, *COM*.

soft keyboard a display resembling the layout of a *keyboard* is presented on a *terminal* screen. A *light pen* is then used to enter *characters* in *machine-readable* form by pointing the pen at each required character in turn.

soft sectoring the identification of *sector* boundaries on a *magnetic disc* by using recorded information. Contrast with *hard sectoring*.

software the instructions, *programs*, or *suite* of programs which are used to direct the operations of a computer, or other *hardware*.

software documentation instructions on how to use a *software package*.

software house a commercial organization which specializes in writing *software* (computer *programs*) for clients.

software package see *package*.

software tools computer *programs* that can write other programs.

SOLINET South-eastern Library Network. A network joining over 200 libraries in the SE United States. It allows them to share *data processing* facilities and access to bibliographic information and resources.

SOM start of message. A *character*, or group of characters, transmitted to indicate that a new message immediately follows.

SOP *standard operating procedure*.

sort 1. any print *character*. 2. to arrange items of information into a sequence according to some rule. Such a rule normally directs that items having characteristics in common should be brought together in the sequence. To make this possible each item is allocated a descriptor which designates its specific characteristics.

Source, The a US interactive *videotex* system which is accessible via *microcomputers* (thus avoiding the regulatory restrictions covering use of television-based videotex). It is supplied by the Telecomputing Corporation of America.

source data acquisition the direct entry of data into a computer at the point where the data originates, eg at a 'point-of-sale' where a cash register is combined with a *terminal* connected to a computer.

source document an original document from which *data* are prepared in a form acceptable to a computer.

source language a term used especially in *machine translation*, *machine-aided translation* and *HAMT*. It refers to the language from which translation is to be made. (See also *target language*.)

space segment a term in satellite communication, referring to the space part of the enterprise, as opposed to the earth, or ground, part. Thus, space segment costs are those of the satellite and its launching, as distinct from those of the ground stations, etc.

Speakeasy a *high level programming language* designed to have as many commands as possible in 'plain English'.

special character a character that is neither a letter, nor a numeral, nor a blank, eg # or $.

specificity indexing of a document is deemed specific if the index terms used are coextensive with the concept to be indexed, and are described precisely. As the degree of matching between indexing and concept lessens, the indexing loses specificity. As a simple example, if the concept to be indexed is 'rat', then 'mammal', 'rodent', 'rat' are terms of increasing specificity.

spectrum in information tyechnology, this refers to the range of *electromagnetic frequencies* available for use in telecommunications. By no means all the available frequencies are currently employed. Diagram overleaf. Most of the range at present in use is shown

in the diagram, which also indicates the names commonly adopted for each band of frequencies. Note that the scale is logarithmic, not linear (ie the frequency changes by a factor of 10 at constant intervals along the scale). Thus the named frequency bands, which occupy roughly equal intervals across the page, are of very different size in terms of the range of frequencies that they occupy (their *bandwidth*). A bandwidth at 'high frequency' is much larger than that at 'medium frequency', which is higher than that at 'low frequency'. The *information-carrying capacity* of a communication channel is roughly proportional to its bandwidth: so the higher bands on the diagram have a much greater capacity.

Transmission in each band of frequencies has its own particular characteristics and problems. The diagram indicates the main uses for each of the bands, both for radio and cable transmission.

Instead of frequency, the electromagnetic spectrum can also be described in terms of the range of wavelengths of the electromagnetic waves. Wavelength is inversely proportional to frequency (ie the higher a wave's frequency, the shorter its wavelength).

speech recognition in the context of information technology, speech recognition means the identification of human speech by electronic means. The basis of speech recognition is the matching of the wave patterns of the speaker with patterns stored in the computer's memory. This is a difficult task, principally because of the complex acoustic wave patterns associated with the human voice. Any particular sound consists of a mixture of many waves with different *frequencies* and *amplitudes*. In addition, the pattern of the sound is only stable for a short time (about ten milliseconds), so that the frequencies and amplitudes continually vary. The sequence of patterns also depends on where in a word the sound appears, and on what sounds precede and follow it. Finally, different people speak differently, and even individual speech varies with time and emotional state.

At present, devices for recognizing speech are effective only within narrow limits. They vary both in the number of words they recognize (their vocabulary) and in

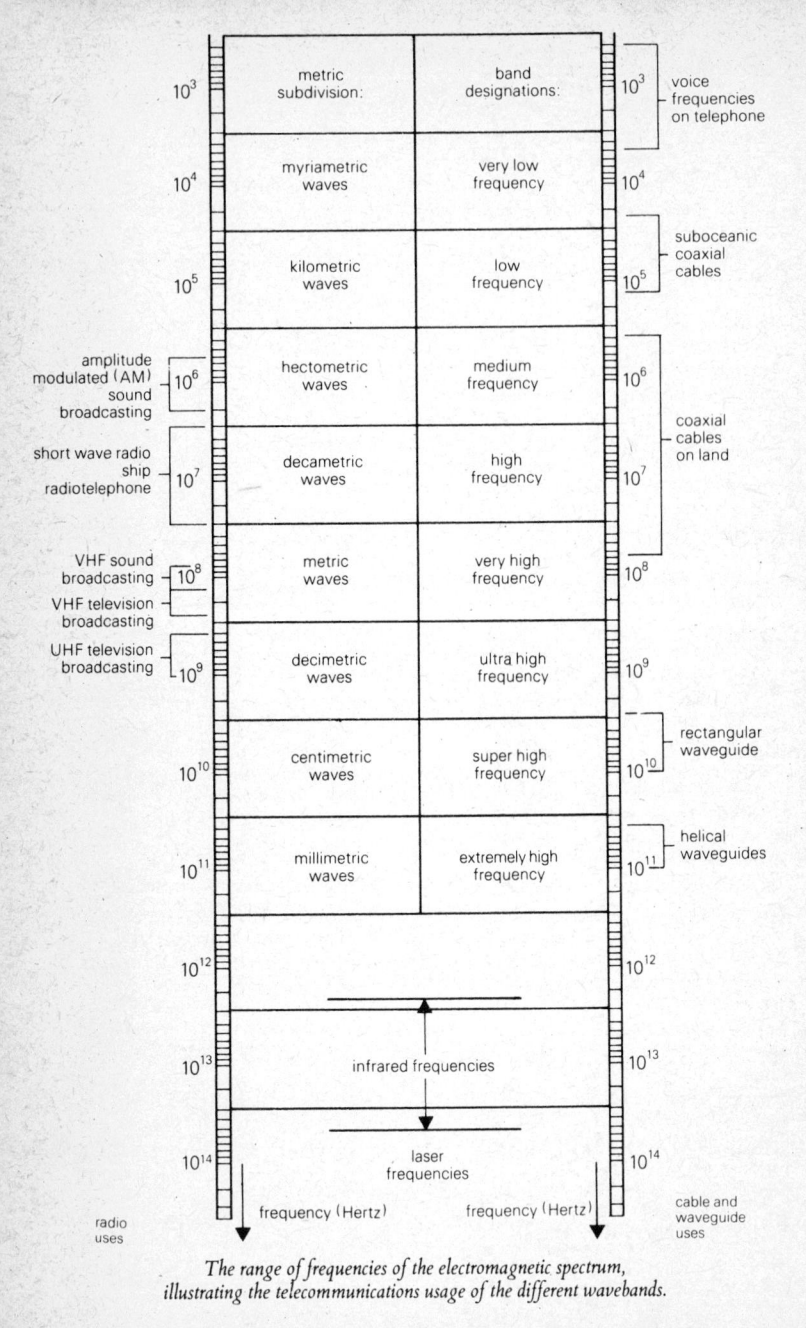

The range of frequencies of the electromagnetic spectrum,
illustrating the telecommunications usage of the different wavebands.

cost. The most sophisticated can recognize some connected-speech (that is, strings of words). Simpler models only recognize one word at a time – they operate in 'discrete word mode'. Currently, a good vocabulary consists of a few hundred words.

The basic steps of speech recognition are as follows. First, some means is used to determine the amplitude spectrum of the incoming voice signal. This is referred to as 'feature extraction', or 'preprocessing', and it can be performed in several ways. The most common is by direct measurement of the spectrum amplitude using a series of *filters*. Another popular technique is *linear predictive coding*, in which the speech signal is represented by the parameters of a filter whose spectrum best fits that of the signal. Features extracted in this way, averaged over perhaps 10 milliseconds, are then sampled say 50 times per second. At this point, or before, the data are *digitized*. Next, the input features are processed to determine the beginnings and ends of sounds, so that the input and reference pattern can be 'time-aligned'. (A technique known as *dynamic programming* is often used here.) This is one of the most difficult tasks in speech recognition. Beginnings and ends of words (ie word-detection) are normally defined by changes in speech energy. The features within the word are then split into 'time slices' for comparison with the reference data. The reference word differing least from the input is the recognized word.

The important performance criteria for speech recognizers are error rate, speaker independence, vocabulary size and ability to recognize connected speech. To some extent these criteria are linked. Speaker-dependent machines (the majority) have to be 'trained': the speaker inputs a series of reference words to the memory, which are stored for later matching. These give some degree of speaker-independence, but the error rate for new speakers is often much higher. The larger the vocabulary to be handled, the greater the possibility of error. For current machines it runs at a few percent.

The main use of the limited speech recognition machines presently available is in the industrial and military worlds. Here, a limited vocabulary set is sufficient for routine operation, eg warehousing, inspection, where the ability to convey information to a computer by voice, whilst simultaneously performing other tasks, is an advantage. Applications considered for the consumer market include voice-controlled television, ovens and wrist-watches.

The effect on information in the office could be extremely significant. Automatic dictation machines are now forecast by the end of the century. Given a good enough vocabulary, these could replace most office keyboarding operations. For simpler tasks, voice operation will come in a good deal earlier. Communication with a computer (ie where the computer replies or initiates a conversation) also requires *speech synthesis*. This is easier, and so is currently more advanced than speech recognition.

speech synthesis the production of speech using artificial means. The synthesis of recognizable speech is simpler than the corresponding task of *speech recognition*, although to make human speech sounds is difficult enough.

Two important methods of speech synthesis are:

i. waveform digitization. This uses the human voice to generate words which are then stored in the computer. The words are spoken into a microphone and the *waveform* of the speech is digitized (ie it is sampled at frequent intervals to produce numbers proportional to the *amplitude* of the wave form). The numbers are stored in *bits*. Speech is synthesized by a *program* which calls up the separate words, puts them together, and then reconverts them to sound (using a *digital-analog(ue) converter*).

ii. *formant* synthesis simulates electronically the sounds of the human voice, rather than using an actual human voice. This is done by generating simple *frequencies* and then modifying them, eg through *filters*, to simulate the complex sounds of a voice.

Speech synthesis has found several applications, eg in toys, and in translation. It is used as an aid to the blind in its developed form, by using *optical character recognition* to scan text, and then reading the text aloud using speech synthesis, eg the *Kurzweil* reading machine.

speech synthesis data capture a method

of using speech as direct *input* (see *voice input*).

Spidel a French *host* (Société pour Informatique) which is linked to *Euronet-Diane*.

SPIN Searchable Physics Information Notes. A *bibliographic database* derived from the publications of the American Institute of Physics. It can be searched via *Lockheed*.

split catalog(ue) a *catalog(ue)* with multiple entries. For example, a *bibliographic database* might contain entries by subject, by author, by year of publication and by title.

split screen the use of a single screen to display separate sets of images. For example, a page of text may be shown in one section of a screen and the record of an *interactive* dialog(ue) with a computer system in another.

SPOOL simultaneous peripheral operation on-line. The *real time* operation of a computer from *on-line* terminals.

spooler *software* which allows *input* and *output* devices to be shared by a large number of users in an orderly fashion and without interference. Spoolers may also control *input* or *output* sequences in accord with a pre-*programmed* specification of priorities.

spooling the use of *spoolers* to achieve effective use of *input/output* devices.

SPS 1. *string process system*. 2. *symbolic program system*.

squeezeout ink method of printing to aid *optical character recognition*. The outline of each *character* is printed darker than its centre.

SRCNET the UK Science and Engineering Research Council *packet switched* data transmission *network*. Researchers in receipt of SERC grants are able to hook into *SRCNET* without any charge.

SRIM *selected research in microfiche*.

SSC *station selection code*.

SSIE Current Research a *database* produced by the Smithsonian Science Information Exchange. It offers multidisciplinary *current awareness* services via *Lockheed* and *SDC*.

STAIRS Storage and Information Retrieval Systems. *IBM software* for *information retrieval*.

stand-alone (capability) the ability of a piece of equipment to operate independently of any other equipment. For example, a *work station* that can operate independently is often called a 'stand-alone system'.

stand-alone system see *stand-alone*.

stand-alone word processor see *word processing* and *stand-alone*.

standard operating procedure a regular, or common, mode of operation, eg of a computer.

standard subroutine a *subroutine*, available to users of a computer, which is designed to solve a well-established problem, or set of problems.

start of text character a *character* used in a transmission to indicate that all headings, eg time of transmission, name of sender and receiver, have been completed, and that the beginning of the message proper follows.

'state of the art' (technology) technology which is currently under development, and will be implemented in the next *generation* of devices.

static RAM *RAM* in which up to one *byte* can be stored at an *address* for as long as power is supplied to a device. (See, in contrast, *dynamic RAM*.)

static video see *still video*.

station 1. a *terminal* or *data processing* facility connected to the computer which the user is using. 2. a *machine location*.

station selection code a computer *code* used to indicate that *access* is required via a particular *station* (1.).

statistical word association see *word/ character frequency techniques*.

Statsionar a series of *communications satellites* launched by the Soviet Union into *geostationary orbit*. Though intended to provide the same sort of coverage as *Intelsat*, the system is mainly used at present to transmit radio and television program(me)s to the Soviet Union. (See also *Molinya*.)

status lines in *word processing*, information displayed to indicate to the operator factors relating to layout, eg spacing, and progress of work, eg line currently being typed.

status word 1. a word communicated by the *computer* to the user to indicate what sort of information should be *input* next, eg *data*, instructions. Examples of such words are 'ready' and 'error'. 2. a word which communicates information about the condition of a *peripheral* unit, eg a warning of some type of malfunction.

stereofiche a form of *microfiche* which allows graphic material to be viewed in three dimensions. The system involves the use of a polarizing stereo box and special viewing glasses in conjunction with a standard *microfiche reader*.

STD subscriber trunk dialling. Direct dialling to a distant location by a telephone subscriber.

STI scientific and technical information.

still video a *telecommunications* technique whereby a telephone is linked to a screen and calls are accompanied, or interspersed, by static images, eg of the caller, or a document. Provision of static images allows a lower *bit rate* than is required for making pictures and/or higher *resolution*.

stop code in *word processing*, an instruction to the *printer* to stop, eg in order to change the *fount*.

stop list in *automatic indexing*, a list of words which the computer is instructed not to use as indexing terms. An index is normally based on substantive terms, so the computer is usually instructed to place words such as 'and', 'but', etc, on the stop list.

storage 1. a *storage device*, or the medium on which information is stored. 2. the process of storing information.

storage capacity the amount of information which a *storage device* can accommodate. Sometimes also called *memory capacity* (see *memory* and *backing storage*).

storage device a device used as *backing storage* for information, eg *magnetic tape* or *magnetic disc*.

storage tube a *CRT* used in some *VDUs* which requires no *refreshing* of the image. Images on such a tube can only be altered by clearing the screen.

store and forward refers to a form of electronic communications, eg *electronic mail*, in which a message is not sent directly to its destination but is stored in a computer *file*. Only a notification of the existence of the message is sent. The message itself is transmitted when the recipient calls for it. (See, in contrast, *messaging*.)

STR *synchronous transmitter-receiver*.

string 1. a group of *items* which are arranged in sequence according to a set of rules. 2. a set of consecutive *characters* in a *memory*.

string process system a *software package* designed for the manipulation of *strings* of *characters*.

stringy floppy magnetic *storage*, consisting of a continuous loop *tape* cartridge.

stroke a mark used to form characters in *optical character recognition*. (See also *key stroke*.)

stroke centreline a line used to designate the midpoint of characters in *optical character recognition*.

stroke edge imaginary lines equidistant from the *stroke centre line* in *optical character recognition*.

stroke edge irregularity deviation of the edge of a *character* from its *stroke edge* in *optical character recognition*.

stunt box a device which performs such functions as *line feed*, carriage return, etc, in a *teleprinter*.

STW the initials stand (in transliterated Chinese) for Experimental Communications Satellite. A planned series of Chinese *communications* satellites in *geostationary orbit*.

STX start of text character. A *control character* which indicates the end of a heading, or introductory message, and the beginning of the actual text.

stylus 1. synonym for *light pen*. 2. device used in conjunction with a *graphics tablet* to *input* and manipulate graphical information (see *graphics tablet* and *computer graphics*). 3. pick-up from a disc, eg *video disc*.

subroutine a part of a *program (routine)* which can be called into operation when required.

subscript a *character* that lies below the normal *baseline*, eg X_2.

subset a modulation/demodulation device used as a communications link (see *modem*).

substitution table a layout chart for a *keyboard*. It is used in *word processing* to show which standard *character* keys can also be used as *special character* keys.

suite (of programs) a number of inter-related *programs* which can be *run* consecutively as a single *job*.

supercomputer a very fast, high-capacity *mainframe* computer, manufactured in small numbers for use where highly complex, rapid calculation is required, eg in weather forecasting.

superfiche *microfiche* with a reduction between 50x to 90x. Between 190 and 400 images can be placed on an A6 sheet. It is one of the forms in which *computer output microform* (COM) can be produced.

superscript a *character* that lies above the normal baseline, eg X^2.

suppress to prevent printing.

SVR super video recorder. A format for *video cassettes* developed by Grundig.

SWAMI software-aided multifont input. *Software* for *OCR* systems.

switching in telecommunications, the means of interconnecting users. Most switching is still carried out by telephone exchanges, but other types of switching exist for special purposes, eg for distributing computer data. (See also *circuit switching*, *message switching* and *packet switching*.)

switching centre (center) a location where multiple *circuits* terminate and incoming messages can be transferred to the appropriate outgoing circuit.

Switch Stream-1 name given by *British Telecom* to its *packet-switched data transmission network*.

Switch Stream-2 a *packet-switched data transmission network* to be introduced in the 1980s by *British Telecom*. It will transmit both *data* and voice messages, using some of the new *System X* exchanges.

syllable in computing, a string of *characters* or part of a *word*.

symbolic address a convenient label for an item or variable in a computer program. It often takes the form of an abbreviation – for example, ATHR for author, or YR for year.

symbolic program system a computer system which accepts *high level programming languages*.

SYNC refers to the synchronizing pulses used to provide a stable reference frame for television pictures.

synchronizing signal a signal which accompanies the transmission of data. The signal can be sent by the station transmitting the data, or from a separate source. In either case, its role is to ensure that the data are transmitted and received in synchronizm with a clock (see *baseband*).

synchronous with a constant time between successive events, eg in the transmission of *bits* or *characters*.

syndetic 1. interconnections within a system. 2. cross-referencing in an *index* or *catalog(ue)* of documents.

synopsis see *synoptic*, which is often used as an equivalent.

synoptic a synoptic is a concise publication in a journal which presents the key ideas and results of a full-length article. It includes an *abstract*, diagrams, references,

etc, and is refereed in the normal manner. The full-length article is either also published, elsewhere or subsequently, or it is made available from a repository.
It has been suggested that a synoptic, rather than an abstract or full-text document, may be the best method of providing information on a *database* intended for *information retrieval*.

synoptic journal a journal which publishes *synoptics* rather than full-length articles.

syntax error a mistake in the formulation of an instruction to a computer.

system an organized set of components (human beings, equipment, etc) which

DOCUMENT
Paper documents and reports of all varieties.

PUNCHED CARD
Punched cards including stubs.

MAGNETIC TAPE

OFF-LINE STORAGE
Offline storage of either paper, cards, magnetic or perforated tape.

ON-LINE KEYBOARD
Information supplied to or by a computer using an on-line device.

CLERICAL/MANUAL OPERATION
A manual operation not requiring mechanical aid.

SORTING, COLLATING
An operation on sorting or collating equipment.

KEYING OPERATION
Operation using a key-drive device.

PERFORATED TAPE
Paper or plastic, chad or chadless.

ON-LINE STORAGE
Described more specifically in the following two symbols.

DRUM STORAGE

DISC STORAGE

COMMUNICATION LINK
Automatic transmission of information from one location to another via communication lines.

AUXILIARY OPERATION
A machine operation supplementing the main processing function.

Selection of symbols used in systems flowcharts

interact in a regulated fashion.

Système International see *SI*.

system-resident usually applied to *software* to indicate the instructions and data which form an integral part of the computer system. (See, in contrast, *media-resident software*.)

systems analysis a technique for analysing systems, and determining the scope for improved efficiency through the introduction of computers and *new information technology*.

systems flowchart a *flowchart* diagram which represents the relationship between 'events' in a *data processing* system, and hence describes the flows of data through, and within, the system. There are various standard sets of symbols used in system flowcharting. One of the most commonly used is shown below. Supplementary program flowchart symbols are often used (see the diagram accompanying the *program flowchart* entry). See diagram on page 169.

systems software see *operating systems*.

System X a computerized telephone switching system developed by *British Telecom*.

SYSTRAN the name derives from a compounding of the words 'Transatlantic System'. It is a *pure MT* (ie fully automated) translation system providing translations between English, French, Russian and Spanish. Initially developed in the US and used by US Government agencies, it has been adopted and developed worldwide. In particular, the European Commission has carried out considerable work on the system. However, extensive practical use of SYSTRAN has not yet occurred, due to the fundamental problems besetting pure MT (see *machine translation*).

tactile keyboard a *keyboard* display which is laid out on a flat surface. A *character* is registered by touching its *key* location lightly with a finger.

tag synonymous with *flag*.

tail a *flag* indicating the end of a *list*.

tape see *paper tape* and *magnetic tape*.

tape comparator a device which automatically compares two supposedly identical *punched tapes*, row by row, and stops when there is a discrepancy.

tape drive a device for winding and unwinding reels of *tape*. Applies especially to *magnetic tape* on a computer.

tape limited when the relatively low speed of the *tape* unit is the limiting factor in determining the rate at which *data processing* takes place.

tape punch a piece of equipment for punching holes in *paper tape*. The information to be punched is normally input from a *keyboard*.

tape reader can apply to any piece of equipment which *reads* tape, though applied originally to *paper tape*.

tape verifier a device which compares a previously *punched tape* with a second manual punching of the data, *character* by character, or row by row.

Target a *pure MAT* system which uses a *terminology bank* to translate scientific and technical documents in English, French, German and Spanish. The system can be operated *on-line* in conjunction with *text-editing* facilities.

target language a term used in *machine translation*, *machine-aided translation* and *HAMT* to refer to the language into which a document is to be translated. (See also *source language*.)

tariff the rate at which charges are made for use of facilities provided by a *common carrier*.

TCM terminal to computer multiplexor (see *terminal*, *computer* and *multiplexor*).

TDM time division multiplexing (see *multiplexing*).

TDMA *time division multiple access*.

teaching machine normally refers to a machine which performs *computer-aided instruction*. (See also *educational technology*.)

TEAM Terminology, Evaluation and Acquisition Method. A German *pure MAT* system based on an *automated dictionary* and a *terminology bank*. It offers *on-line* interrogation of its *databases*, giving *target language* equivalents for any *input* terms, in eight major European languages. In addition, it is able to produce a variety of *off-line* translation aids and services (see *machine-aided translation*).

technische kommunikation the German term for *information technology*.

TeD see *Teldec*.

Teldec form of *video disc* developed by Telefunken and Decca.

Telecom Australia the Australian *telecommunications* agency.

telecommunications the transmission and reception of *data* in the form of electromagnetic *signals*, using broadcast radio or transmission lines. However, it is often used for almost any transmission of signals.

teleconferencing a general term for any conferencing system employing telecommunications links as an integral part of the system. There are two main types – *computer conferencing* and *video conferencing*. Hybrid configurations are also possible. Teleconferencing may be accompanied by other facilities, such as *electronic document delivery*, to enhance its effectiveness.

telecopier a device for *facsimile transmission*.

Teledata a Norwegian *viewdata* (interactive *videotex*) system.

T

telefax the linking of photocopying machines for the transmission of images.

Telefax the name of French and West German *facsimile transmission* services.

Telefax 201 a Dutch *facsimile transmission* service.

telegraph a method of transmitting electrical signals, using simple on-off conventions to provide the code.

teleinformatics a term used to refer to data transfer via telecommunication systems.

Telemail a US *electronic mail* service.

telematics information technology and its applications. Derived from the French 'télématique'.

télématique the French term for information technology.

telemetry the remote measurement of physical quantities, eg electrical quantities, fluid flows.

Telenet a US *packet-switched* telecommunications network.

teleordering a system which automates the way in which booksellers order from publishers. Taking the UK system as an example, participating bookshops and publishers each have a small *intelligent terminal*. During the day, the bookshop terminal is used to collect and store orders. Books are uniquely identified via their *International Standard Book Number* (ISBN). The terminal applies procedures to the orders, and checks that no necessary data are missing. At the end of the working day, the terminal is switched to standby. During the night, using public telecommunications (telephone) lines at cheap rates, a central *minicomputer* system automatically dials the terminal at each bookseller in turn. The terminal responds with details of the collected orders to the central minicomputer, which stores them on *magnetic tape*. The tape is then transferred to a *mainframe computer* for further processing.

The data from the orders are matched with the publishers' information and orders are put in a format suitable for each publisher. The minicomputer system then returns a confirmation (or error indication) for every order to each bookshop terminal. These data are stored in the terminal memory until the next day, when they can be printed out. At the same time, the minicomputer sends a file of orders for relevant books to the terminal of each participating publisher. Publishers not participating in the system receive their orders by post.

This ordering system is currently limited to booksellers and book suppliers (publishers, wholesalers and distributors) in the UK. The diagram shows the main features of the system.

Other countries have teleordering systems at various stages of development. For example, in the US there is a system, called ACCESS, developed by a wholesaler, Ingram, in conjunction with Software Sciences Limited (who designed the UK system). A library supplier (Baker and Taylor) has produced systems called LIBRIS and BATAB; whilst the Bowker Company had instituted a Computerized Acquisition System (CAS), based on 'American Books in Print' as the database. Another library supplier (Brodart) has introduced a teleordering service for the New York area. In West Germany, teleordering was introduced in the mid-1970s by two large wholesalers with incompatible systems: BESSY (Bestell-System) from KNO (Koch, Neff, Oetinger/Köhler Volckmar) and the Litos TBA (Telefonischer Bestell-Abruf) of Libri (Lingenbrink). Denmark also has a well-established system called Bookseller Data (ABD).

Telepack a US *telecommunications* service offering combined voice and data *channels*.

telephone data set a unit used to connect a data *terminal* to a telephone circuit (see Modem).

telephotography the name in the US for the transmission of news pictures using *facsimile transmission*.

teleprinter a device resembling a type-

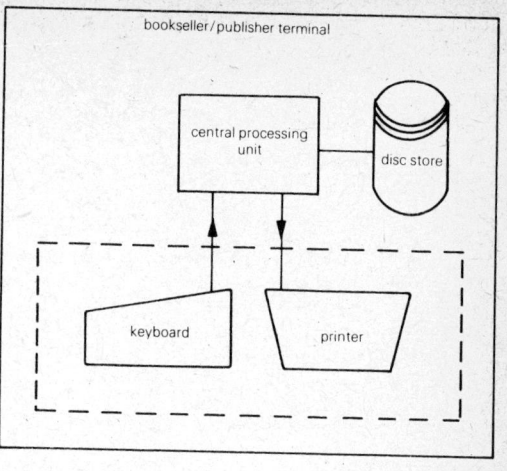

The British teleordering system, used by booksellers to place orders with publishers and other suppliers, showing the data flow. Below is a schematic diagram of one of the system's terminals.

bookseller/publisher terminal

central processing unit

disc store

keyboard

printer

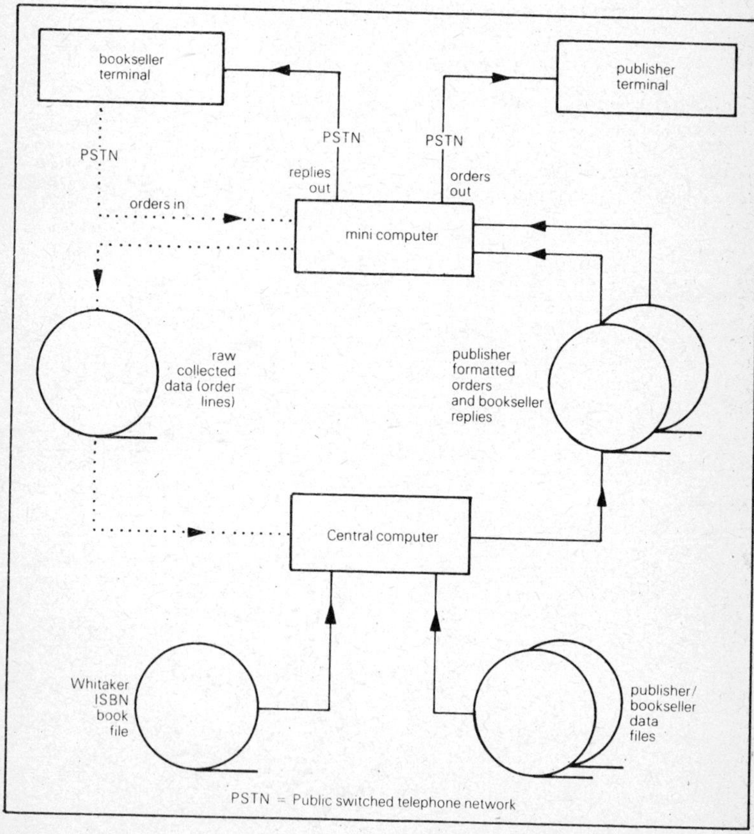

bookseller terminal

publisher terminal

PSTN

PSTN   PSTN

replies out

orders out

orders in

mini computer

raw collected data (order lines)

publisher formatted orders and bookseller replies

Central computer

Whitaker ISBN book file

publisher/ bookseller data files

PSTN = Public switched telephone network

writer which is connected to a *telegraphic* circuit. It can be used to transmit, or receive, and print out data.

teleprocessing *data processing* and transmission using computers and telecommunication *channels*.

Teleset a Finnish *viewdata* (interactive *videotex*) system.

teleshopping the use of interactive *videotex* to select goods for purchase and to place orders.

telesoftware the transmission of computer *software* using *videotex*.

Télésystème a French *host* system.

Teletel the brand name of the French *Antiope-Titan videotex* system.

teletex the *CCITT* name for a system which transmits data between *terminals*. Its ultimate purpose is to link *word processors* via the public telephone network. It represents a combination of *text* editing with high speed telex-related equipment. If manufacturers produce word processors which conform to teletex standards, current problems of *incompatibility* between word processors could be overcome. Teletex will then not be dependent on a single *dedicated/network* in the way that telex is.

teletext see broadcast *videotex*.

teletype grade a level of circuit suitable for *telegraphic communication*.

teletypesetting a system for remote *typesetting* first used before the Second World War. The typesetter is controlled by *punched páper tape* input, either directly, or over a communications link.

teletypewriter normally called a *teleprinter*.

telex a world-wide *telegraphic* service, established by *Western Union*, which permits interconnection between *teleprinters*. Operators use terminals to produce *paper tape*, which is then fed into a reader for transmission over a *dedicated/network*.

The telex system is relatively old and slow. However, it is so well established that it will take some time to replace. It is likely to be superseded by *teletex*, or some related communication system.

Telidon a Canadian *viewdata* (interactive *videotex*) system.

temporary storage in computer *programs*, this refers to *storage locations* reserved for intermediate results.

tera- a prefix denoting one million million (10^{12}).

terminal a device for sending and/or receiving data over a communications channel. It usually has both a keyboard and a *VDU*. The term is sometimes used to describe a single *workstation*, eg of a *word processor*, whether or not it possesses a communications capability.

terminal transparency the ability, in a telecommunications network, to allow incompatible terminals to communicate by automatic *code conversion* and line control conversion.

terminology bank a computer-based glossary of terms which provides translations for each entry. It is used in *machine-aided translation* systems, such as *Termium*, *Target* and *Eurodicautom*.

Termium a Canadian *pure MAT* system which uses a *terminology bank* with a *permutation index* (allowing access to expressions via any significant word they contain). Termium is the largest and oldest terminology bank in the world, and offers *on-line machine-aided translation* in English and French throughout Canada.

Tex a *suite* of *routines* (devised by Knuth) to enable technical text (including mathematics) to be input at an ordinary computer *terminal*.

text editing the editing of text on a computer. It may be carried out on any form of computer, from a *mainframe* with appropriate *software* to a dedicated *word processor*.

Textline an international business *database* covering the major UK, French and German newspapers, as well as Japanese newstape. Sources are collated and condensed into *abstracts* which are indexed and loaded daily for *on-line* access.

text management used as an equivalent term for *text processing*.

text processing the computer editing and subsequent production of textual material. It is often used as a synonym for *word processing* and for text management (including systems for storing and retrieving text). The distinction in usage from *word processing* lies primarily in the amount of data handled: working with very large quantities of text is usually referred to as 'text handling'.

text processing system a computerized system designed to *edit* and manipulates text, which is then output in a specified medium and *format*.

text reader-processor a device which combines a text-reading capability (usually by *optical character recognition*) with a *data processing* facility (usually a *microprocessor*). Data can thus be *read* and processed into a form suitable either for a particular use, or for transmission to a particular piece of equipment.

text retrieval system a computerized system which allows the retrieval of documents included in its *database*(s) by selecting words, word groups and sequences in the documents themselves (see *information retrieval system* and *information retrieval techniques*).

text structure input a method of *input* for *on-line searching* of a *database* containing details of chemical structures. It resembles *graphic structure input*, except that the description of a chemical structure is given by the user using commands selected from a *menu*. (See also *chemical structure retrieval*.)

thermal printing a method of printing using paper coated with heat-reactive dyes that darken at temperatures between 100°C and 150°C. A matrix of small elements, in

good thermal contact with the paper, is selectively heated to form individual characters.

thermography thermography is used in conjunction with *letterpress* printing. Slow-drying ink is used and, after printing, the paper is dusted with a resinous powder. Heat is then applied and the ink fuses and swells giving an embossed appearance.

thesaurus a structured collection of terms which is used to index documents. It was originally used to control *post-coordinate* systems, but can also provide a subject headings list.
A thesaurus typically provides control of synonyms, and indicates how a particular *index term* is related to others in the *indexing language*. This capability can help define terms when carrying out a *search* during *information retrieval*.

thimble printer a type of *printer* used in some *word-processing* systems. The name derives from the shape of the *fount*.

thin film (memory) see *magnetic film*.

thin window display a (limited) form of electronic display; typically containing up to 96 characters, and often a single line only. It is used in some *word-processing* devices.

Thomson-CSF optical video disc a *video disc* system developed by Thomson-CSF. The approach taken is basically similar to that of *Philips/MCA Discovision*; but the discs are thinner and more flexible. They are also therefore transparent: both sides can be read without turning the disc over, simply by altering the focus of the reading *laser* (see video disc).

thread a group of *beads* strung together for testing purposes.

throughput the amount of work processed during a specified time period.

tie line a private communications channel (usually leased from a *common carrier*) connecting two, or more, *private branch exchanges*. Sometimes also called a tie trunk.

tie trunk see *tie line*.

time division multiple access a technique for dividing a stream of signals between *terminals* in a network by allocating time slots.

time division multiplexing see *multiplexing*.

time sharing simultaneous *real time* use of a computer by a number of users. It is made possible by ultra-high-speed *switching* of *processing* time.

TIP TOP tape input/tape output. Describes a system which uses *magnetic tape* as both *input* and *output* medium.

token a distinguishable unit in a set of *characters*.

toll 1. a charge for linking to a *network* of telephone lines beyond a prescribed boundary. 2. US name for UK *trunk*.

toll office a US exchange which controls the switching of *toll* calls. (See also *trunk exchange*.)

toll switching trunk a line connecting a toll office to a local exchange. (The US name for UK *trunk junction*.)

tone the relative shades of dark and light in an image (see *tone illustration* and *half tone*).

tone illustration an illustration in which there is a continuous variation in *tone*, as in an ordinary photograph. (Contrast with *half tone*.)

Topic the electronic information system of the London Stock Exchange. *Pages* can be retrieved from the Exchange's own *database*, *Epic*, and displayed, in colour, on *terminals*. Price rises are marked in blue and falls in red, allowing the experienced user to scan the screen very rapidly to determine how a particular sector of the market is fairing. Topic can be accessed outside London via leased *British Telecom* lines.

touch screen terminal a *terminal* with a screen which is sensitive to touch. The positions touched (usually by a finger) are recorded by the *computer* as coded signals. Data can thus be input simply by touching the screen. (For related devices, see *soft keyboard*, *tactile keyboard*, and *light pen*).

TOXBACK Toxicology Information Back-up. In the *BLAISE* implementation of *TOXLINE*, most of the *National Library of Medicine* files are searchable *on-line*. For the remaining files, TOXBACK offers a back-up service: a query can be entered on-line, but the search is processed over-night.

TOXLINE Toxicology Information On-line. A *database* produced by the US *National Library of Medicine* covering toxicology, environmental pollutants, drug chemistry, pharmacology, etc. It is accessible via *BLAISE*, *DIMDI* and *NLM*.

track a path along which a sequence of signals can be impressed on a recording medium. For example, a *magnetic tape* can have several tracks in parallel.

traffic the messages which pass through a communications system.

transaction any event which requires that a *record* be *processed*, eg updating a *file*.

transactional videotex the use of interactive *videotex* to effect transactions, eg place orders, make bookings.

transborder data flow refers to the flow of data, or information, across national boundaries. Owing to regulations imposed by governments, *PTTs*, etc, transborder flow is often more restricted than communication within national borders.

transceiver a *terminal* which can both transmit and receive information.

transcribe to copy data from one *storage medium* to another, with, or without, some form of translation.

transducer at the most general level, a device for converting energy from one form to another. In information technology, it is an *input*, or *output*, device designed to convert signals from one *medium* to another. For example, a loudspeaker is a transducer which converts electrical signals into acoustic signals.

transfer operation the movement of data from one *storage* area, or *medium*, to another.

transfer rate the speed at which data can be transferred from one *peripheral* device to another, or from a *CPU* to a peripheral.

transistor the name derives from *transfer of* electricity across a re*sistor*. An electronic device made of semi-conducting material which can be used to amplify or switch signals (see *semiconductor*).

translate has a particular significance in computer *displays*: it means to adjust the position of an image on the screen.

transliteration a representation of the characters of one alphabet by those of another, usually corresponding approximately in sound.

translucent screen synonym for *opaque screen*.

transmission the electrical, or electromagnetic, transfer of energy. In information technology, it usually refers to the transfer of data signals from one point to another.

transmission loss a decrease in *signal* power during *transmission*. (Note that it does not mean the loss of a transmission.)

transnational data flow see *transborder data flow*.

Transpac French *packet-switching* telecommunications network.

transparent 1. refers to any computer activity that goes on unseen by the operator. 2. refers to the combination of data and *word processing software* into a single *program*.

transponder a device which both receives and transmits data. A received signal can be amplified and retransmitted at a different frequency. *Communication satellites* usually have more than one transponder. Retrans-

mission must take place at a different frequency, since otherwise the powerful retransmitted signal will interfere with the weak, incoming signal.

transverse scanning a form of scanning in which the head moves across, rather than along, the recording tape. In particular, it is a technique used in some *video tape recorders*.

TRIAL technique for retrieving information from abstracts of literature. A system for the manipulation of *bibliographic data* in the form of *abstracts*.

trunk a telecommunications channel which carries signals between *switching* centres, often over long distances.

trunk exchange a UK exchange which controls the switching of *trunk* calls. (See also *toll office*.)

trunk junction a line connecting a *trunk exchange* to a local exchange. (The UK name for US *toll switching trunk*.)

truth table a form of *Boolean operation table* in which the two possible values for each *operand* and result are 'true' or 'false' (see *Boolean algebra* and *Boolean operation table*).

TS *time sharing*.

TSI *test structure input*.

TSS 1. time shared system. 2. time sharing system (see *time sharing*.)

TTL transistor-transistor logic. A technology for constructing *integrated circuits* from *transistors* formed on a single slice of silicon. Lends itself to the production of circuits with a high operating speed, but is unsuited to the most complex types of circuit.

TTS 1. *teletypesetting*. 2. teletypesetting code: a development of *Baudot code*.

TTY *teletypewriter*.

tube refers, in origin, to the *cathode ray tube*, but is often used nowadays as a synonym for *screen*. Used in the US, formerly, as the equivalent term for the UK 'valve'.

Tulsa a *database* produced by the University of Tulsa (based on Petroleum Abstracts) and made available via *SDC*.

turn-around time 1. the time required to reverse the direction of a transmission. 2. the time between submitting a *job* to a computer and getting back the results.

turnkey system 1. a complete computerized system installed by a single supplier, who takes total responsibility for the production, installation and operation of all the *hardware* and *software*. 2. a system in which the user does not need to know any programming or *operating system* commands, eg loading programs. When turned on, the system goes into a preprogrammed activity, eg *word processing* or *information retrieval*.

TV-SAT/TFD a Franco-German *communications satellite* project set up in 1980. The plans are for a *direct transmission satellite* to provide television program(me)s in Western Europe. TV-SAT refers to the German component, and TDF to the French component. (See also *L-SAT*.)

TVT television typewriter. A device for displaying data on a television screen.

TWX Teletypewriter Exchange Service. A public *switched teletypewriter* service offered by *AT&T* in the US and Canada.

TX *telex*.

Tyme-gram a US *electronic mail* service operated by Tymnet Inc.

type face 1. the design, or style, of *characters* produced by a particular *printer*. 2. the printing surface of a piece of type which bears the character to be printed.

type fount (font) a complete collection of *characters* of a particular size and design. (See also *fount*.)

typographic quality a synonym for *graphic arts quality*.

UA *user area.*

UCS *universal character set.*

UDC *universal decimal classification.*

UHF ultra high frequency: from 300 to 3000 megaHertz. (Compare with *VHF.*)

UKITO an acronym for the United Kingdom Information Technology Organisation. A group of British companies involved in the application of information technology.

ULSI ultra large scale integration. Similar to *VLSI*; but, while VLSI refers to *microprocessors* composed of several tens of thousands of components, ULSI refers to microprocessors with approximately 100,000 components. (See also *LSI.*)

ultrafiche a *microfiche* with such small images that 3000 pages can be mounted on one 4-inch x 6-inch fiche.

Ultra LSI another way of writing *ULSI.*

ultrasonic sound with a *frequency* too high to be heard by the human ear (above 20 *kilocycles* per second).

ultrasonics technology involving the use of *ultrasonic* waves.

ultraviolet electromagnetic radiation in the range between visible light and x-rays: of the order of 10^{15}-10^{16} cycles per second (see *spectrum*).

U-matic a form of *video cassette* developed by Sony.

umbrella information providers in *viewdata*, an *information provider* who provides services to other information providers.

UMF ultra microfiche: also called *ultrafiche.*

UMI ultra microfiche (see *ultrafiche*).

unbundling the separation of charges for *hardware* and *software* in the buying and selling of computers and software.

uncontrolled language see *information retrieval techniques.*

UNESCO United Nations Educational, Scientific and Cultural Organization.

UNISIST 1. United Nations Information System in Science and Technology. A UNESCO program(me), designed to bring about a world information network. Mainly a coordinating body. One product of relevance to information technology is the *International Serials Data System* (ISDS). 2. Universal System for Information in Science and Technology: a classification system.

UNISTAR User Network for Information Storage Transfer, Acquisition and Retrieval (see *information retrieval system*).

uniterm indexing a system of *indexing* which uses a single term to describe documents. These single terms (uniterms) are then used to retrieve documents (see *information retrieval systems*).

universal character set a facility on a *printer* which permits any standard *typeface* to be chosen when printing a document.

universal decimal classification a *decimal classification system* developed by *FID.*

Universal Postal Union a United Nations agency responsible for promoting international cooperation between national postal services.

universal product code each product, eg in a retail store, is described by a *code* in the form of a series of bars of varying widths. This is usually physically attached to the product.
The code can be read by an *optical scanner* at the point and time of purchase. The information can then be used to produce an automatic receipt and to record transactions. This aids the retailer with stock control, compilation of sales figures, etc. etc (see *bar code*). (For an alternative, see *magnetic stripe systems*.)

unjustified text which has an irregular right-hand margin.

U

up has a specific meaning in the context of computer systems. Such a system is described as 'up' when it is operating, and 'down' when it is not.

UPC *universal product code.*

updatable microfiche by combining micro-imaging processes (see *microform*) with *photocopying*, it is possible to produce a specially coated *microfiche* to which images can be added as and when required.

update to amend the *records* in a *file*, usually to take account of new information.

uplink the *earth station*, and its transmitted signals, to a *communications satellite*.

upper case capital letters of a fount, eg 'A' as opposed to 'a'.

UPS uninterrupted power supply. A protection against breakdowns and blackouts.

uptime the time during which a computer system is '*up*' (ie operating). 'Uptime' is often expressed as a percentage, and this is used as an index of reliability; eg '99.5 per cent uptime' – an indication of a reliable system.

UPU *Universal Postal Union.*

upwards compatibility describes the situation when a *program* written for one computer can also be run on a second computer, though a program written for the second computer will not run on the first (see *compatibility*).

URL User Requirements Language: a *high level programming language*.

USART universal synchronous/asynchronous receiver/transmitter. A device which can receive and transmit *data* in the form of *synchronous* or *asynchronous* signals.

USASCII United States of America Standard Code for Information Interchange (see *ASCII*).

USDA/CRIS US Department of Agriculture/Current Research Information System. The USDA compiles a *database* on current agriculture-related research projects, and makes it accessible via *Lockheed*.

user area the area on a *magnetic disc* in which a user's semi-permanent *data*, or *programs*, can be stored.

user-friendly a system with which relatively untrained users can interact easily. This normally implies the use of a *high level programming language*, and often of graphical representation.

user interest profile a definition of the information needs of an individual expressed as a series of *index terms* (see, eg *selective dissemination of information*).

Userkit a device for improving and simplifying access to *on-line information retrieval systems*. The userkit enables *search statements* to be prepared before going *on-line*, thus saving time, and also gives access to international systems using stored *host* addresses, *passwords* and *log-on* sequences. The Userkit is a box, containing a *microprocessor*, which plugs in between the user's *terminal* and the *modem* (see *on-line searching*).

USITA United States Independent Telephone Association. A US-based organization, but having members (independent telephone companies) in many countries. It deals with common problems, technical standards and regulatory matters.

USPO United States Post Office.

USRT universal synchronous receiver/transmitter. A device which can receive and transmit *data* in the form of *synchronous* signals.

UV *ultra violet.*

V volt: *SI* unit of electrical force.

VAB voice answer back. A pre-recorded voice response used in conjunction with a telephone-type *terminal* connected to a computer.

VACC *value added common carrier*.

validity the extent to which repeated applications of a process obtain the same result.

value added common carrier a *common carrier* which does not itself establish telecommunications links, but which leases links from other carriers. It can thus create a computer-controlled network offering specific telecommunications services.

VAN value added network.

variable field see *field*.

variplotter a high-precision graphic recording device (see *plotter*).

VCC Video Compact Cassette. A *video cassette recorder* developed by Philips.

VCR a *video cassette* system, developed by Philips, now replaced by *V2000* format. The initials are also used more generally as an abbreviation for *video cassette (recorder)*.

VDI 1. visual display input. 2. video display input. The *input* at a *VDU*.

VDT 1. visual display terminal. Originally this meant an *on-line* display, but it is often used as a synonym for a *VDU*. 2. less frequently, the initials may stand for video display terminal.

VDU visual display unit. A device equipped with a *cathode ray tube* for the visual display of information. Usually connected to a *keyboard* for inputting and editing information. (See also *VDT*.)

vector graphics a form of *computer graphics* utilizing *line drawing displays*. Lines are entered on a display screen and then manipulated using a *keyboard* or *light pen*. Vector graphics are extensively used in *computer-*

aided design. (See, in contrast, *raster graphics*.)

Venn diagram a diagram showing the relationship between sets of items: more particularly, the nature of their overlap. For example, the Venn diagram below shows the nature of overlap between different types of terminology. 1. Computer terminology. 2. Library terminology. 3. Telecommunications terminology. 4. New information technology. John Venn, after whom the diagram is named, was a 19th Century logician. (See also *Boolean algebra*.)

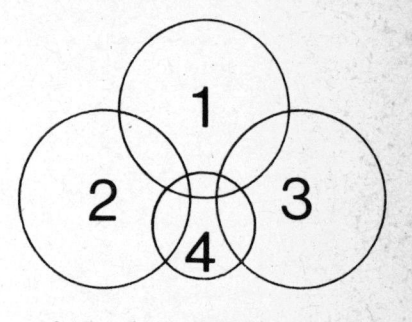

verification the checking of *coded* data against original source data to detect errors.

vertical raster count the number of vertical divisions in a *raster*. (See also *raster count*.)

very large scale integration (VLSI) electronic circuits constructed on a single *chip* with a complexity equivalent to over ten thousand *transistors* and up to a hundred thousand *transistors*. (See also *LSI* and *ULSI*.)

VHD *Video High Density*.

VHF very high frequency: from 30 to 300 *megaHertz*. (Compare with *UHF*.)

VHS Video Home System. A *video cassette* system developed by JVC.

video a general term implying a visual display. Sometimes used as a synonym for a *VDU*.

video bandwidth the maximum rate at

V

which dots of illumination, eg *phosphor dots*, can be displayed on a screen.

video camera a camera which records images on *magnetic tape* for playback, eg on a *video tape recorder*.

video cassette (recorder) a device for visual recording (especially of television program(me)s) onto *magnetic tape* contained in a plastic housing (see *videotape recorder*).

video cassette journal subscribers periodically receive a *video cassette* and accompanying booklet. The cassette can be played by using a *videotape recorder* attached to a television set. Costs are kept down by the subscriber returning the cassette for reprogramming prior to the next 'issue' of the journal.

video conferencing a form of *teleconferencing* where participants see, as well as hear, other participants at remote locations. *Video telephones* can be used for a limited form of video conferencing over the *public switched telephone network*. However, most such systems will be on *leased lines*, or lines that can carry full-quality television pictures. Most existing systems operate between cities, eg by the provision of suitable studios in the major centres, either for public hire or for private use. A typical arrangement is shown in the diagram.

video disc a disc, typically made of plastic, containing recorded visual and sound information designed for play back on a television screen. The main recent investment in video discs has been for launch on the home entertainment market, as a competitor for *video cassette recorders* (VCRs). In general, video discs have two main advantages: i. the materials from which they are made are less expensive than those for VCRs, or conventional film; ii. the disc itself is light and compact and can be very easily stored and transported. Some, but not all, types of disc have the further advantage of *random access* to various parts of the disc (in contrast to tapes in video cassettes). A disadvantage is that the discs in systems currently available cannot record information, but only play-back pre-recorded material.

The video disc need not be used solely for the storage of sound and pictures. It has the capacity to store large amounts of text, or mixtures of text, sound, graphics and moving pictures. This, together with random access and other capabilities of some systems, has led to experimental investigations of the institutional uses of video discs for education and for information storage,

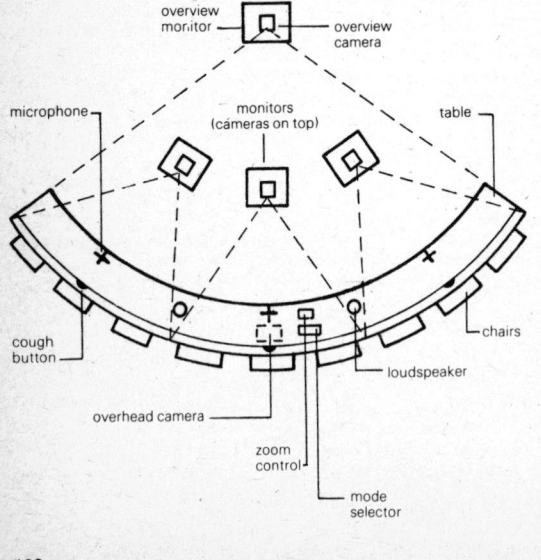

Video conferencing layout, allowing transmission between distant conference areas using ordinary television facilities. The 'mode selector' has a 'normal mode' position in which cameras switch automatically to the person speaking.

overview monitor
overview camera
microphone
monitors (cameras on top)
table
cough button
chairs
overhead camera
loudspeaker
zoom control
mode selector

retrieval and transfer.

There are two main types of video disc:
a. contact video discs; b. non-contact (optical) video discs. It is the latter which are considered to be of the greater general importance in information technology; while a third type – the optical digital disc – is likely to be important in mass storage of information.

a. *contact video discs:* early types followed methods somewhat similar to gramophone recordings, but the main system today is 'capacitive' (ie the electrical signal to be read from the disc is determined by its ability to store electricity at each point). Although it would in theory be possible to design a capacitive system where the *stylus* and disc did not come into contact, it would in practice be very expensive. For the system to act as a capacitor the video disc must contain conductive material and it, or the stylus, must be coated with non-conductive material so that the disc and the stylus do not 'short' out on contact. There are currently two main makes of this type of system. RCA make a *Selectavision* system; whilst JVC make a similar capacitative system, *VHD*.

b. *optical video discs:* two of the main organizations, Philips and MCA, have agreed on some common standards for their systems (see *Philips/MCA Discovision*). Again more than one system exists, but the basic principle is always that the tracks on the disc are monitored by optical *laser* beams. The other major optical system comes from Thomson-CSF of France (see *Thomson-CSF optical video disc*). This is again a laser-based disc, but uses a transparent disc, rather than the reflective disc of Philips/MCA.

In most information, as opposed to entertainment, applications, the industrial format optical disc is generally regarded as the most suitable (because of its ability to present still pictures, hold frames of text and provide rapid access to any frame). Video discs can be used for storage of television pictures and sound, sound only, or for *digital* storage of data, or all of these in combination. Taking the Philips/MCA disc as an example, this has 54,000 frames per side. But each television frame is fairly limited in the amount of information it can store. For example, the characters on one average typewritten A4 page would require

some eight normal television frames (ie 6,000-7,000 pages per side of a disc). Single frames cannot therefore be used to store complete pages of text for viewing on domestic television receiver screens. In the UK these have 625 lines (525 in the US) which would not give the required *resolution*. High-definition monitor screens of 2000 lines are available which would give much better definition; using special 'scan conversion' techniques, normal commercial video screens could be used to play on these monitors.

However, as just mentioned, the video disc can be used to store information in digital form. In this case, the storage of a single side is approximately 10^{10} *bits*, or roughly 1 million pages of 1250 characters, which is vastly more text than if it is used in a direct, television-compatible way. An index for each frame will indicate whether digital or video information is being carried on that frame, and the disc output can be directed to a video monitor or a computer for subsequent output. Thus the video disc could have considerable application as a storage and retrieval device, as well as a full-text store and publishing medium, by virtue of its cheap storage capacity, easy access, and ability to mix text and graphics.

c. *optical digital discs:* optical video discs have the digital information encoded on a standard video signal. An alternative approach is that of the optical digital disc, where digital information is placed on a disc directly by a user with a computer and the appropriate equipment. The data can be read directly after they are written onto the disc. This is called the direct read after write (*DRAW*) process. The storage capacity of these discs is likely to be between 10^{10} and 10^{11} bits. Currently they are being developed as a *mass storage* device for computers. They may be grouped into disc packs of 6 discs. As such, their storage capacity could be 10^{12} bits, compared with a conventional magnetic disc pack of comparable physical volume, storing 3×10^9 bits.

Whereas errors on video discs used for video may be noticeable, they seldom cause problems in receiving the visual or sound image. Similar error rates on digital information could be very important however, since they could lead to failure to retrieve information, etc. The monitoring and replication processes of optical video discs introduce a number of errors, which may present a considerable source of difficulty for their use in digital *information storage and retrieval*.

videogram a pre-recorded *videotape*, or *videocassette*.

Video High Density a *video disc* system developed by JVC. Like *Selectavision* of RCA, this is a *capacitive* system. A *laser* is used to record the signals on a glass master disc. The discs produced from this do not have grooves: the *stylus* is guided along the appropriate track by a special signal.

Video Patsearch a UK service (offered by Pergamon) for searching US patents. The index to, and text of, patents is held on computer *files*, whilst the *graphics* are recorded on *video discs*. The files can be searched from a user's *terminal*, and both text and graphics can be accessed (see *on-line searching*).

video recorder see *video tape recorder*.

video tape recorder a device which, when connected to a television set, can be used to record both sound and pictures on *magnetic tape*. This tape is normally enclosed in a cartridge (a *video cassette*). Video tape recorders can also be used to 'play back' these tapes, whether they have been prepared in this way (ie from a television broadcast or transmission), or in some

other way, eg using a *video camera*. Also called a *video recorder* and *video cassette recorder*.

video telephone a telephone which also transmits and, depending on the system, receives an image on a screen. (See also *video conferencing*.)

videotex videotex is a generic term referring to any electronic system that makes computer-based information available via *VDUs*, or appropriately adapted television sets, to a dispersed and reasonably numerous audience. Videotex systems can be divided into two main categories.
a. broadcast videotex: where information is carried from the computer to the receiver by radio waves. The role of the user in this type of system is generally restricted to the choice of one from a limited number of *pages* for display on a screen, using a simple push-button selection device. For example, the two systems of this type in operation in the UK – *Ceefax* and *Oracle* – broadcast (without charge to the viewer) frequently up-dated information on such topics as weather forecasts, sports results, etc. This information is 'piggy-backed' on existing television transmissions, which limits the number of pages available to a few hundred.
b. interactive videotex: where the information is carried from the computer to the receiver by cable (usually telephone lines). Use of cables leads to two advantages over broadcast videotex. The amount of information available is limited by the capabili-

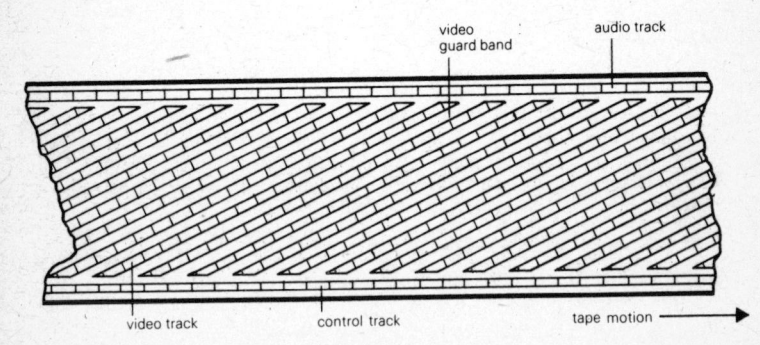

Typical track layout of video tape

ties of the computer, rather than by the mode of transmission. Thus, the UK form of interactive videotex – called *Prestel* – already has over a quarter of a million pages accessible. Instead of passively receiving information, users can interact individually with the computer.

In making Prestel available, the British *PTT* acts as a *common carrier*. The material to be transmitted is fed in by independent *information providers*, who hope to cover their costs by charging users for consulting pages. The approach varies from country to country. For example, one of the main current developments in France is the government-backed distribution of VDUs to telephone users to replace their print-on-paper telephone directories. Apart from such public videotex systems, private systems are also coming into existence: they are usually aimed at restricted audiences in the business sector.

Owing to the large number of pages on offer, *information retrieval techniques* are necessarily more complex for cable-based, than for radio transmission. Optimum methods by which inexperienced users can track down the information they need are still being investigated. Similarly, the presentation of material on the screen (and especially *graphics*) is becoming more sophisticated. These developments are important for the future growth of the interactive capabilities of videotex. Planned or existing user interaction includes carrying out numerical calculations, playing games, consulting sales catalogues and ordering goods *on-line*, recording votes in local referendums and exchanging messages with other subscribers to the network (a form of *electronic mail*).

One problem with videotex is the different generic names that have been applied to it. Thus 'interactive videotex' is sometimes also called 'viewdata', whilst 'broadcast videotex' may be referred to as 'teletext'. The latter term must not be confused with '*teletex*'.

Videotext a West German *teletext* system. The term is sometimes used as a synonym for *videotex*.

vidicon (tube) the basic element of a video

Diagram of the two basic videotex reception systems. Interactive videotex receives information by cable, whereas broadcast videotex receives it by radio waves. Videotex also covers the reception of local signals from a home microcomputer. For a fuller illustration of interactive videotex, see Prestel.

camera (normally only black and white).

Viditel a Dutch *viewdata* (interactive *videotex*) system.

Vidon a Canadian *viewdata* (interactive *videotex*) system.

viewdata an alternative term for interactive *videotex*.

virgin medium a *storage* medium with no data recorded on it.

virtual storage the linking of the main *memory* of a computer with external *storage* in such a way that they can function as one memory of larger capacity.

Visc an early *video disc* system, which employed a *stylus*.

viscose film film which is rendered photosensitive by impregnation with a light-sensitive dye. It is sometimes used for *microforms*.

visual display unit see *VDU*.

VLF very low frequency: a frequency below 30,000 *Hz* (see *spectrum*).

VLP Video Long Player. A Philips/MCA video disc system.

VLSI *very large scale integration.*

vocoder a device for *speech synthesis* which transmits sufficient information for the voice to be synthesized, without attempting to preserve the original voice waveform. It aims at semantic clarity, but not at copying the voice. This means that the sounds produced will not necessarily resemble the voice of the original speaker.

VOGAD voice operated gain-adjustment device. A device which reduces fluctuations in the level of reproduced speech, so improving the *signal-to-noise ratio*.

voice activation the operation of any device that can be trained to respond to the human voice.

voice answer back a system in which a computer gives responses (to a user's commands) in the form of prerecorded voice messages.

voice band refers to a *bandwidth* of the order of 4,000 Hz. (See, in contrast, *narrowband* and *wideband*.)

voice grade channel a channel suitable for transmission of speech. It must therefore cover a range of sound frequencies from 300-3400 Hertz (cycles per second).

voice input see *direct voice input*.

voice mail see *voice message system*.

voice message system an electronic system for transmitting and storing voice messages, which can be accessed later by the person to whom they are addressed.

voice notes a facility for *terminal* users who cannot, or will not, type to add voice messages to a computerized text. A user can inspect a document on a *VDU* and append audible messages ('voice notes') to it via a microphone plugged into the terminal.

voice recognition see *speech recognition*.

voice synthesis see *speech synthesis*.

voice unit a measure of the average *amplitude* of an electrical signal which represents speech. Also called a volume unit.

volatile memory *memory* which loses its *information* when power is cut off. It is to be contrasted with *permanent memory*.

volume unit synonym for *voice unit*.

Votes a *database* which presents the voting records of members of the US Congress on bills, resolutions, treaties etc. It is updated twice monthly, and is accessible via *SDC*.

VSMF visual search on microfilm (see *microfilm*).

VTR 1. *video tape recorder*. 2. video tape recording.

V2000 a *video cassette recorder* system developed by Philips.

VU 1. *voice unit.* 2. *volume unit.* (The two terms are synonymous.)

W Watt: *SI* unit of power.

WACK 'Wait before transmitting positive acknowledgement': a signal sent by a receiving *station* to indicate that it is temporarily not ready to accept a message. (See also *ACK* and *NACK*.)

WADS wide area data service. A data transmission service which operates on a *network* much like that serving a *WATS* telephone service.

wafer a thin slice of silicon which forms the basis of a *chip*.

wait list synonymous with *queue*.

wand a device, shaped like a stick, which can be used to recognize *optically coded labels*.

WARC *World Administration Radio Conference*.

watermark magnetics a system for encoding information onto a *magnetic stripe*. It is used for magnetic stripes fixed to plastic credit and banking cards. (For their use, see *automatic bank teller machines*.) The watermark's distinctive feature is its performance. The encoded information is not erased, or altered, if the stripe passes through a strong magnetic field.

WATS wide area telephone service. A service which allows a subscriber to make local calls at a flat monthly rate (ie no charge for individual calls).

waveform digitization the conversion of a wave into *digital* form by the generation of numbers proportional to the *amplitude* of the waveform at frequent intervals. In particular, waveform digitization is the name given to one technique of *speech synthesis*.

waveguide a metal tube used for the transmission of very high frequency electromagnetic signals.
Rectangular waveguides are typically used for the connection between *microwave antennas* (aerials) and associated equipment. They are not normally used for transmission over distances of more than a few hundred metres.
Circular waveguides can transmit much higher frequencies than rectangular waveguides (typically 40 to 110 *GHz*). These would be heavily *attenuated* if transmitted through the atmosphere. A helical waveguide is of a circular cross-section, but has copper wire wound round the inside in a helix. This helps *attenuate* undesired modes of the signal.
Circular and helical waveguides are used for transmission over many kilometres. The waveguides must not bend sharply, but can change direction by gradual curvature.

wavelength see *spectrum*.

wavelength multiplexing a technique for transmitting separate signals simultaneously by using a different *wavelength* for each signal. Applies particularly to the transmission of light wavelengths via *optical fibres*.

WB Weber: *SI* unit of magnetic flux.

WDC *World Data Centre*. A number of these were established to promote the collection and international exchange of scientific data. There are three main components of WDC: *WDC-A* in the US, *WDC-B* in the USSR and *WDC-C* in several Western European countries, Australia and Japan. The centres collect data covering the Earth and space sciences.

WDC-A see *WDC*.

WDC-B see *WDC*.

WDC-C see *WDC*.

web in printing, a method of printing onto a continuous length of paper which moves through the machine.

W

weighting in *information storage and retrieval*, the practice of assigning weights to *search terms* to indicate their importance. A number of search terms, eg *descriptors*, may be used, each assigned different weights. Only those items are retrieved which carry descriptors whose combined 'weight' is above some specified threshold.

Westar a series of *communications satellites*

owned by *Western Union* for communication in the United States.

Western Union Corporation a US *common carrier* offering a variety of voice, video and data transmission services, some of which are based on its *Westar* satellite system.

WESTLAW West Publishing Law System. A US *legal retrieval* system. It enables searches to be made for words and phrases in full legal texts, eg of judicial rulings, US codes.

wheel (printer) a *printer* which has its type face mounted on the rim of a wheel. (See also *daisy wheel printer*.)

whirley bird slang for *magnetic disc* equipment.

white space skid a feature of some *facsimile transmission* machines. It allows transmission to be speeded up by the scanner skipping the blank spaces on the document to be transmitted.

wide area data service see *WADS*.

wide area telephone service see *WATS*.

wideband refers to a *bandwidth* of the order of 48,000 *Hz*. (See, in contrast, *narrowband* and *voiceband*.)

Winchester disc a very common type of *hard disc*.

wireless terminal a portable, or handheld, *terminal* that can communicate with a computer by radio.

WISE World Information Systems Exchange. An arrangement between several hundred institutions world-wide to encourage collaboration in, and the exchange of data about, *information technology*.

Wiswesser line notation a widely accepted method of representing the chemical structure of a compound by a *string* of *characters* (see chemical structure retrieval).

WLN *Wiswesser line notation.*

word a group of *characters* representing a unit of data, and occupying one *storage* location. It is an indication of the number of *bits* that can be handled by the *CPU* of a computer in each computing step. Thus it has become a measure of the size of a computer, eg a computer may be described as '32-bit word computer'.

word association see *word/character frequency techniques*.

word/character frequency techniques some *information retrieval techniques* are based on the characteristics of *natural language*. For example, natural language has considerable redundancy, as can be seen from the way garbled messages can often be reconstructed accurately. This provides an opportunity for text compression when putting material into *machine-readable* form. Thus, words do not all occur with the same frequency, nor do letters. In a piece of text,

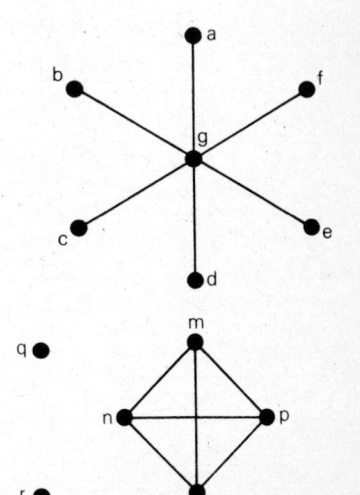

'Star' word (top) association map. The words a to f co-occur strongly with word g, but not with each other.

'Clique' word (bottom) association map. The words m to p co-occur strongly with each other, but not at all with q or r.

typically a small proportion of the words make up a large proportion of the text, while many words are used very infrequently. It is possible to design codes so that a frequent word or character is represented by a code of very short length, thus making savings in computer storage space. One example is the *Huffman code*. Other methods take whole portions of text which may occur frequently, eg phrases, or ends and beginnings of words, and code them. The frequency of occurrence of words in natural language can also be used as an aid in indexing, classification and retrieval. For example, considerable research has been done on the frequency of co-occurrence of words within documents. Co-occurring words can be used as a method for classifying documents. Word association maps can be drawn, some of which can themselves be categorized, eg 'stars' or 'cliques' or 'clumps', giving rise to different types of classification of associated documents (see diagram). The technique is known as 'statistical word association', and originally derived from early studies on machine-aided translation.

word processing is the term generally used in English to describe the computer handling of text. Although it is applied to a number of different functions, the essential elements are the enhancement of typewriting by the addition of *memory*, electronic *intelligence*, and an automated *printer*.

The origins of word processors are to be found in automated typewriters. These used paper tape, which recorded text as punched holes when the typewriter was operated. The punched tape was then re-run through the typewriter to produce further, identical copies of the text. In 1964, *IBM* introduced a much more useful form of this typewriter employing *magnetic tape* as a recording medium. The magnetic tape could be erased and re-used. More importantly, copy recorded on it could be edited. A major limitation of this system was that a typist could not be sure during the course of operations that amendments were being made correctly. This could only be ensured by scanning the final product. The incorporation of a display device overcame this: both text and amendments could be seen before the print-out stage. The first

'screen-based' word processors became commercially available in the early 1970s. Most displays now show at least a major fraction of a standard page of text. All word processors have five elements in common:

i. a *keyboard*. This is usually similar to the QWERTY keyboard of a conventional typewriter, but has a number of additional 'control' keys which are used to perform editing and other functions.

ii. *internal memory or storage*. This is the 'working area' of the word processor. Text is entered, either from the keyboard, or from external storage, and is then available for editing or revision. The larger the internal memory, the greater the amount of text that can be manipulated at one time.

iii. *external storage*. This refers to the storage of text after its creation, or editing. Some systems have magnetic cards, suitable for the storage of letters, but the most common form today is the *floppy disc* or *diskette*. This is available in a range of sizes and storage capacities. It can be removed from the word processor and stored separately. In the future, advances in storage technology may make removable external storage unnecessary.

iv. *logic*. The logic part of the system provides for the editing functions, the means of storing and retrieving text, and, in general, the control of the system. Some of the logic will be in the form of electronic circuits, or *hardware*: but, increasingly, much of the logic is in *software*, or computer *programs*, stored in the internal memory. These programs are brought into action via the keyboard. They drive a *microprocessor* which executes their instructions.

v. *printer*. The printer, or print-unit, may be attached to the keyboard, or (more commonly) may be physically separate. The most common types are *golf-ball* and *daisy-wheel printers*.

These five elements can be combined in various ways, leading to three main types of word processor: *stand-alone*, *shared-logic* and *distributed-logic*. The terminology is not yet completely standardized. For example, distributed logic systems are sometimes called *shared-resource systems*.

Stand-Alone. These are self-contained, single keyboard systems.

Shared Logic and Distributed Logic (Shared Resource). With these systems, a number of

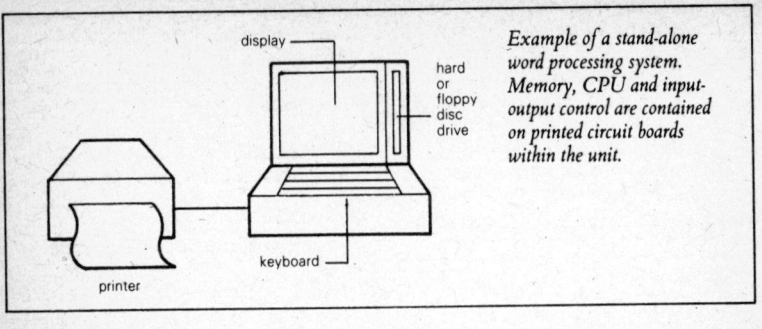

Example of a stand-alone word processing system. Memory, CPU and input-output control are contained on printed circuit boards within the unit.

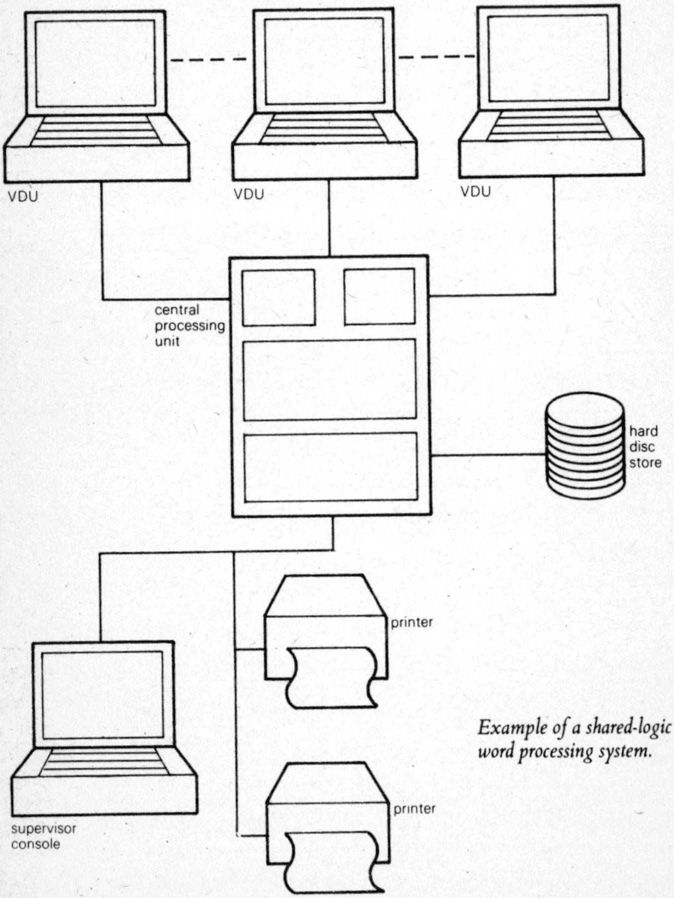

display

hard or floppy disc drive

keyboard

printer

VDU   VDU   VDU

central processing unit

hard disc store

printer

printer

Example of a shared-logic word processing system.

supervisor console

keyboards can share the processing, storage, printing, or other facilities used in conjunction with a word processor. In a shared-logic system, the individual entry points or *work-stations* make use of a central processor; whereas, in a distributed-logic system, each has its own *processor* and uses other shared facilities, eg printers, independently.

Time-sharing computer services allow for a further type of system. Here suitable word processors link with a *main-frame* computer to edit copy. Such a computer has the necessary storage capacity for work with very large amounts of text. Large computers, traditionally involved in numerical operations, are increasingly incorporating text-handling capabilities.

A current trend is for the incorporation of more *logic*, or *intelligence*, into each *work-station* of a word processor. This increases their power and the range of their functions. Thus some word processors can be used for limited data processing, *information retrieval* and *telecommunications* purposes. They can also be linked to other *data capture* devices, eg *OCR scanners* for the *input* of text, or *phototypesetters* to produce high-quality *output*. (See also *text processing*.)

word processor see *word processing*.
See diagram opposite.

word time the length of time required to move a *word* from one *storage* location to another.

word-wrap a *word processing* term. It refers to the way in which a partially typed word is moved to a new line if its length proves too much to fit into the existing line.

work area synonymous with *working storage*.

work file a *file* on *magnetic disc* containing details of the *program* which is currently being run through the computer.

working space synonymous with *working storage*.

working storage temporary *storage* reserved for *data* which are actually being processed.

work station sometimes referred to as an *electronic office*. A work station offers access to such facilities as *telephone*, *telex*, *facsimile*, *personal computer*, *word processor*, *data transmission terminal*, *viewdata* and perhaps even *videophone*, from one integrated unit. Often used in a restricted sense to mean a *terminal* with access to one of these facilities. (See also *station*.)

World Administration Radio Conference international meetings held at intervals and mainly concerned with the allocation of frequencies. Meetings throughout the 1970s concentrated on *satellite communications*. Any frequency allocation system agreed at a WARC would be run by the *International Telecommunications Union*.

World Aluminium Abstracts a *bibliographic database* produced by the American Society for Metals and covering ore processing, metallurgy and end uses of aluminium. The database is accessible via *ESA-IRS*, *Lockheed* and *QL*.

World Patents Index see *WPI*.

World Textiles a *bibliographic database* covering the science, technology, technical economics, management and trade of textile and related industries. The database is compiled in the UK and made accessible via *Lockheed*.

wow and flutter a change in *output frequency* of a signal due to variations in *tape* speeds. Wow applies to slow speeds and flutter to high. Flutter has an additional meaning in connection with video tape (see *flutter*).

WPI World Patents Index. A *database* containing patent information from all the major R&D nations, accessible *on-line* via *Pergamon-Infoline*.

wpm words per minute. A measure of speed of transmission in *telegraph* systems.

wrap see *word wrap*.

wrap-around a way of displaying data on a *terminal* screen. When the screen has been filled (ie a *character* has been entered in the last vacant position at the bottom right hand corner of the screen), then following characters are displayed in the first positions on the screen (working from the top left hand corner), overwriting any characters already present.

write to record *data*, or copy it from one *storage device* to another.

write head an electromagnetic device used to *write* on a magnetic *storage* medium.

write rate the maximum rate at which the phosphor *dot* on a *terminal* screen can produce a satisfactory image.

writing head synonymous with *write head*.

writing tube a *tube* on which an electron beam *writes*, or scans for information.

WS abbreviation for *working storage*, or *working space*.

xerography xerography is a technique whereby an electronic image of a document is formed onto a drum. The drum picks up a black powder by electrostatic attraction. The powder is deposited onto blank paper, and fused into the paper by baking to form the final reproduction of the document.

xerox the trade name for a type of document copying machine, based on *xero-*graphy. Owing to the popularity of these machines, *photocopying* is sometimes referred to as 'xeroxing'.

x-height the height of a *lower-case* letter when *ascenders* and *descenders* are excluded.

x-y plotter synonymous with a *plotter*, also sometimes called a *dataplotter*.

yoke a group of *read/write heads* fastened together.

zap in computing, to erase, especially from programmable read only memory (see *ROM* and *PROM*).

zatacode a *code* used to index records in a *data processing* system.

zero access storage *storage* which gives *access* in a negligible time (see *rapid access memory*).

zero compression the elimination before storage of non-significant zeros (ie those to the left of all digits in a number). For example, the number 00503.010, after zero compression, would be 503.010.

zero elimination the elimination before storage of non-significant zeros (see *zero compression*).

zeroize 1. to fill a storage space with zeros. 2. to reset a mechanical register to zero.

zero suppression synonymous with *zero elimination*.

Appendices

The material in Appendix 1 is from BS 3939: Section 22: 1969 and reproduced by permission of the British Standards Institution, 2 Park Street, London, W1A 2BS from whom complete copies can be obtained.

The following graphical symbols accord with British Standard 3939: Section 22 and are suitable for use with telecommunications diagrams. Although in use outside the United Kingdom, they do not necessarily conform with standards in other countries.

no.	description	symbol

WAVEFORM OR FREQUENCY CHANGERS

no.	description	symbol
22.3.7	frequency multiplier	
22.3.8	frequency divider	
22.3.9	pulse inverter	
22.3.10	code converter from 5-unit binary code to 7-unit binary code	
22.3.11	changer giving clock-time indication in 5-unit binary code	
22.3.12	differentiator	

TELEGRAPH EQUIPMENTS

no.	description	symbol
22.9.1	telegraph apparatus	
22.9.2	telegraph transmitting apparatus	

Appendix 1

no	description	symbol
22.9.3	telegraph receiving apparatus	
22.9.4	telegraph transmitting and receiving apparatus, 2-way simplex	
22.9.5	telegraph transmitting and receiving apparatus, duplex	

NOTE. The T in the general symbols may be shown above the square or be replaced by one or more of the following qualifying symbols to specify a particular type of apparatus

qualifying symbols (to be used with the general symbols 22.9.1 to 22.9.5)

no	description	symbol
22.9.11	tape printing	
22.9.12	perforated tape	
22.9.13	simultaneous printing on and perforating of one tape	
22.9.14	page printing	
22.9.15	typewriter keyboard	
22.9.16	facsimile	

no.	description	symbol

examples of symbols for telegraph apparatus

no.	description	symbol
22.9.21	tape-printing receiver	
22.9.22	tape-printing receiver with typewriter keyboard transmitter (teleprinter)	
22.9.23	printing reperforator (receiving perforator)	
22.9.24	page-printing receiver	
22.9.25	page-printing receiver with typewriter keyboard transmitter (teleprinter)	
22.9.26	page-printing and tape-reperforating receiver with typewriter keyboard transmitter	
22.9.27	facsimile receiver	
22.9.28	automatic transmitter using perforated tape	
22.9.29	typewriter keyboard perforator	
22.9.30	separate reperforator and automatic transmitter	

NOTE. If the tapes are cut and fed one by one to the transmitter, the dashed line between the blocks is omitted.

no	description	symbol
22.9.31	combined reperforator and automatic transmitter with continuous tape feed	

examples of symbols for telegraph repeaters

NOTE. In the following symbols the $\pm$ sign indicates double current and the $\frac{+}{o}$, $\frac{+}{o}$, $\frac{o}{-}$, or $\frac{-}{o}$ signs indicate single current

22.10.11	double current telegraph repeater for duplex working	
22.10.12	double current/single current telegraph repeater for one-way simplex working	
22.10.13	double current/alternating current telegraph repeater	
22.10.14	regenerative telegraph repeater	
22.10.1	telegraph repeater for one-way simplex working: general symbol	
22.10.2	telegraph repeater for two-way simplex working: general symbol	

no.	description	symbol

TELEPHONE SETS

22.11.1 telephone set: general symbol

NOTE 1. It is recommended that these symbols be drawn with the proportions shown
NOTE 2. Qualifying symbols may be added as shown in the following examples:

22.11.2 local battery telephone set

22.11.3 common battery telephone set

22.11.4 telephone set with dial
NOTE. The dots in the circle may be omitted if no confusion can result

22.11.5 telephone set with push-button dialling

22.11.6 telephone set with key(s) or push-button(s) for special facilities other than dialling

22.11.7 telephone set with two or more lines (exchange and/or extension lines)

22.11.8 telephone coinbox set, with or without separate coin-box

no.	description	symbol
22.11.9	telephone set with ringing generator, eg magneto	
22.11.10	loudspeaker telephone set	
22.11.11	telephone set with amplifier	
22.11.12	sound-powered telephone set	

EXCHANGE EQUIPMENTS

no.	description	symbol
22.12.1	exchange equipment: general symbol NOTE. Qualifying symbols may be added as shown in the following examples:	
22.12.2	automatic telephone exchange equipment	
22.12.3	automatic telex exchange equipment	
22.12.4	relay equipment of an exchange or of part of an exchange	
22.12.5	manual switchboard: general symbol	

Appendix 2 Conversion codes

The *ASCII code* and character set.

most significant bits

				column	0	1	2	3	4	5	6	7
			b_7		0	0	0	0	1	1	1	1
			b_6		0	0	1	1	0	0	1	1
			b_5		0	1	0	1	0	1	0	1
bits b_4	b_3	b_2	b_1	row								
0	0	0	0	0	NUL	DLE	SP	0	@	P	`	p
0	0	0	1	1	SOH	DC1	!	1	A	Q	a	q
0	0	1	0	2	STX	DC2	"	2	B	R	b	r
0	0	1	1	3	ETX	DC3	#	3	C	S	c	s
0	1	0	0	4	EOT	DC4	$	4	D	T	d	t
0	1	0	1	5	ENQ	NAK	%	5	E	U	e	u
0	1	1	0	6	ACK	SYN	&	6	F	V	f	v
0	1	1	1	7	BEL	ETB	'	7	G	W	g	w
1	0	0	0	8	BS	CAN	(	8	H	X	h	x
1	0	0	1	9	HT	EM	)	9	I	Y	i	y
1	0	1	0	10	LF	SUB	★	:	J	Z	j	z
1	0	1	1	11	VT	ESC	+	;	K	[	k	{
1	1	0	0	12	FF	FS	,	<	L	\	l	¦
1	1	0	1	13	CR	GS	—	=	M	]	m	}
1	1	1	0	14	SO	RS	.	>	N	^	n	~
1	1	1	1	15	SI	US	/	?	O	—	o	DEL

least significant bits

Key to abbreviations

NUL	null	DLE	data link escape (CC)
SOH	start of heading (CC)	DCl	device control 1
STX	start of text (CC)	DC2	device control 2
ETX	end of text (CC)	DC3	device control 3
EOT	end of transmission (CC)	DC4	device control 4 (Stop)
ENQ	enquiry (CC)	NAK	negative acknowledge (CC)
ACK	acknowledge (CC)	SYN	synchronous idle (CC)
BEL	bell (audible or attention signal)	ETB	end of transmission block (CC)
BS	backspace (FE)	CAN	cancel
HT	horizontal tabulation (punched card skip) (FE)	EM	end of medium
		SUB	substitute
LF	line feed (FE)	ESC	escape
VT	vertical tabulation (FE)	FS	file separator (IS)
FF	form feed (FE)	GS	group separator (IS)
CR	carriage return (FE)	RS	record separator (IS)
SO	shift out	US	unit separator (IS)
SI	shift in	DEL	delete

CC: communication control
FE: format effector

IS: information separator

Appendix 2

Conversion table for *EBCDIC* and *binary* codes.

Binary	EBCDIC	*Binary*	EBCDIC	*Binary*	EBCDIC	*Binary*	EBCDIC
1000 0001	a	1001 0101	n	1100 0001	A	1101 0101	N
1000 0010	b	1001 0110	o	1100 0010	B	1101 0110	O
1000 0011	c	1001 0111	p	1100 0011	C	1101 0111	P
1000 0100	d	1001 1000	q	1100 0100	D	1101 1000	Q
1000 0101	e	1001 1001	r	1100 0101	E	1101 1001	R
1000 0110	f	1010 0010	s	1100 0110	F	1110 0010	S
1000 0111	g	1010 0011	t	1100 0111	G	1110 0011	T
1000 1000	h	1010 0100	u	1100 1000	H	1110 0100	U
1000 1001	i	1010 0101	v	1100 1001	I	1110 0101	V
1001 0001	j	1010 0110	w	1101 0001	J	1110 0110	W
1001 0010	k	1010 0111	x	1101 0010	K	1110 0111	X
1001 0011	l	1010 1000	y	1101 0011	L	1110 1000	Y
1001 0100	m	1010 1001	z	1101 0100	M	1110 1001	Z

About the Authors

A. J. Meadows is Head of the Astronomy and History of Science Departments and Project Head of the Primary Communications Research Centre at the University of Leicester. He is a member of various committees, including the Royal Society Scientific Information Committee and has written extensively on new technology and information science.

M. Gordon has carried out research into a variety of aspects of the communication of scientific information, and is a Research Associate at the Centre.

A. Singleton has worked in publishing and as a product engineer in the electronics industry. He is a member of the Board of Scientometrics and is now a Research Fellow at the Centre.